ENERGY
THE MASTER RESOURCE

An Introduction to the History, Technology, Economics, and Public Policy of Energy

ROBERT L. BRADLEY, JR.

RICHARD W. FULMER

KENDALL/HUNT PUBLISHING COMPANY
4050 Westmark Drive   Dubuque, Iowa 52002

Photos on pages 2, 16, 62, 80, 116, 142 and 176 from Corbis.
"Earth's City Lights" cover photo from NASA.
Back cover photo of authors courtesy of Temple Webber Photography.
Remaining cover images © 2004 PhotoDisc, Inc.

Copyright © 2004 by Robert L. Bradley, Jr.

ISBN 0-7575-1169-4

Printed in the United States of America
10 9 8 7 6 5 4 3 2

To our next generation . . .

Catherine and Robert Bradley III
Elizabeth, Katherine, James, and Marie Fulmer

EPIGRAPHS

❝Coal, in truth, stands not beside but entirely above all other commodities. It is the material energy of the country—the universal aid—the factor in everything we do. With coal almost any feat is possible or easy; without it we are thrown back in the laborious poverty of early times.❞

— **William Stanley Jevons, *founder of mineral economics* (1865)**

❝Energy is the master resource, because energy enables us to convert one material into another. As natural scientists continue to learn more about the transformation of materials from one form to another with the aid of energy, energy will be even more important. . . . For example, low energy costs would enable people to create enormous quantities of useful land. . . . Reduction in energy cost would make water desalination feasible, and irrigated farming would follow in many areas that are now deserts. . . . Another example: If energy costs were low enough, all kinds of raw materials could be mined from the sea.❞

— **Julian Simon, *economist/optimist* (1996)**

❝Affordable energy in ample quantities is the lifeblood of the industrial societies and a prerequisite for the economic development of the others.❞

— **John Holdren, *environmentalist/pessimist* (2001)**

ABOUT THE AUTHORS

Robert L. Bradley, Jr. is president of the Institute for Energy Research (IER) and a senior research fellow at the University of Houston. He is author of five books on energy, including the 2-volume *Oil, Gas & Government: The U.S. Experience*. Richard W. Fulmer is senior fellow at IER and a systems analyst in the energy industry.

ACKNOWLEDGMENTS

The authors would like to thank the following readers for their review of the entire manuscript: Peter Grossman (Butler University), Gürcan Gülen (Institute for Energy, Law & Enterprise, University of Houston), Howard Hayden (University of Connecticut, emeritus), John Jennrich (Federal Energy Regulatory Commission), William Leffler (Shell Oil Company, retired), and Al Mayo (independent consultant).

Individuals who reviewed sections of the manuscript or provided important information include Gene Callahan (author, *Economics for Real People*), John Christy (University of Alabama at Huntsville), Indur Goklany (American Enterprise Institute), Ken Green (Fraser Institute), Joel Schwartz (American Enterprise Institute), and Christopher Namovicz and Tom Petersik (Energy Information Administration, U.S. Department of Energy).

The authors alone are responsible for any omissions or errors of fact or analysis.

CONTENTS

PREFACE

Energy is *the master resource*. It is essential for life and provides comfort and protection against the elements. People in developed countries can scarcely imagine living without lighting, conditioned air, indoor plumbing, electric ovens, mechanized transportation, medical devices, computers and all the other elements of the modern, energy-intensive world.

Yet energy is at the center of many concerns and controversies. Much of the world's oil supply is concentrated in the politically unstable Middle East. Oil, natural gas, and coal—the basis of today's economy—cannot be reproduced like most things we use. They are nonrenewable resources that can be consumed only until their reserves are depleted. Air pollution in our cities is an unwanted by-product of the fuels that power our cars and generate most of our electricity. Combustion of fossil fuels releases greenhouse gases that may alter the climate in undesirable ways. Oil spills around the world kill sea life and foul beaches.

Plentiful energy has enabled the great population growth of the past two hundred years. But can enough energy be produced cleanly enough to support the population rise—from six billion to nine billion people—that is predicted later this century? Is energy, or, more accurately, human ingenuity up to the challenge?

Energy and energy policy have provoked many emotional debates. A particularly vitriolic controversy sprang up between a Danish statistician, Bjørn Lomborg, and a group of natural scientists. Lomborg presented a mountain of facts in his book *The Skeptical Environmentalist* (Cambridge University Press, 2001) and concluded that alarm concerning energy supply and the climate is unwarranted. *The Scientific American* published a four-part rebuttal that claimed that Lomborg was not only wrong but also *incompetent*. To emphasize this point, the magazine subtitled the article, "Science defends itself against The Skeptical Environmentalist." The debate quickly escalated into an international brawl over issues of science, scholarship, professional ethics, and even censorship. Ideas and facts presented in our book will touch on many of these issues.

❖　❖　❖

This primer on the history, technology, economics, and public policy of energy explains what energy is and how its use has evolved over the centuries.

It also discusses the "sustainability" of the modern energy economy from the standpoint of both available resources and energy's effect on the environment. The book combines introductory information that might be found in an energy encyclopedia with the topics that have put energy at the forefront of the national consciousness. For the general reader, information is presented in a nontechnical way so that little prior knowledge of the field is required. For the more advanced reader, the leading writings on each side of the debate are referenced to facilitate further investigation.

❖ ❖ ❖

The opening chapter presents the basics, including an historical overview to help the reader understand where we are and how we got here. Wind, solar, and biomass (wood, agricultural byproducts, etc.), which for centuries were the world's primary sources of energy, no longer make significant contributions in the modern world. Carbon-based energies—the oil, coal, and natural gas that comprise approximately 85 percent of United States and world consumption[1]—ended the renewable age because they were far more concentrated, portable, reliable, and cost-effective energy carriers.

Large central-station electricity plants replaced distributed generation (then called "isolated plants") because of their greater efficiency and lower cost. Electric cars dominated the automobile market for a short time but were displaced by vehicles powered by internal combustion engines. Ethanol, an agricultural product, was a viable, even popular, fuel a century ago, but it was displaced by fossil fuels. The reasons for the switch to gasoline and diesel still remain relevant.

There are reasons to appreciate our energy past and to be optimistic about our energy future. As this book documents, carbon-related technologies are doing well in a two-front war against resource depletion and pollution. Economists believe that the fossil-fuel resource base will be adequate for many decades and probably centuries. Technological improvements and capital turnover (that is, replacement of older vehicles, machines, and power plants with newer, more efficient equipment) promise to continue to make our air and water cleaner in the decades ahead even as energy consumption increases.

The issue of energy sustainability now centers on the effects of fossil fuel extraction and usage on global warming—a subject that is central to this book. The science surrounding climate change in general and global warming in particular is complex, and many uncertainties remain. Crucial issues are still being hotly debated, including what the effects are on the climate after

---

[1]These percentages are based on *measured* energy use, which does not include such things as burning wood in a fireplace or using a window to capture sunlight.

150 years of rising atmospheric greenhouse gas concentrations and whether the risks of anthropogenic climate change are outweighed by the risks of climate change *policy*.

Of course, this book does not have all the answers—the science and economics of climate change are not advanced enough yet to provide them. But given the history of energy outlined in these pages, there can be confidence that any arising problems will inspire the human ingenuity and policy reform necessary to solve them. As the late resource economist Julian Simon emphasized, human ingenuity is the *ultimate resource* that, when applied to the *master resource* of energy, can enable people to enjoy longer, more comfortable, and more productive lives.

❖   ❖   ❖

This book offers a basic, readable introduction to energy for advanced high school students, college and graduate students, policymakers, and the interested public. Far too much reporting on information about energy and environmental issues is simplistic, agenda-driven, and alarmist. Without historical context—and without appreciation for basic economic principles such as opportunity cost, cost/benefit analysis, and decision-making in the face of uncertainty—problems can be seen where none exist, and "solutions" can create real problems where none existed before. If this book can help to balance the debate in the classroom, the office, and at home, the authors will have achieved their goal.

ENERGY
THE MASTER RESOURCE

An Introduction to the History, Technology, Economics, and Public Policy of Energy

THE BASICS

1

INTRODUCTION

In 1798, British philosopher, economist, and clergyman Thomas Robert Malthus predicted that human population would outpace food production. Population, he said, expands geometrically (e.g., 1, 2, 4, 8, 16,. . .), while agricultural production grows only arithmetically (e.g., 1, 2, 3, 4, 5,. . .). Ultimately, only disease and starvation could keep the human population in check.[2] Small wonder that economics became known as "the dismal science."

But the predicted disasters never came. Instead, food and natural resources became more plentiful. Technology so increased human productivity that, in the West at least, food and resources take an ever-smaller portion of family income.

Some argue, however, that Malthus's predictions were not wrong, just premature. The Earth's food and energy supplies will eventually run out, they say, unless humanity is first overcome by pollution. In the 1970s, it appeared to many that these dire warnings were finally coming true. Gasoline shortages caused long and frustrating lines at service stations, and tight natural gas supplies disrupted home and business life. The environment was suffering from decades of neglect. Many lakes and rivers were dead or dying, and choking smog was common in major American cities.

Yet, the energy crisis ended by the early 1980s, and within ten years the oil shortage was replaced by an oil glut. Fish returned to rivers and lakes long thought dead. By 2000, the air in Los Angeles no longer burned visitors' lungs. In Europe and North America, population growth either stopped or declined.

In 2001, however, California experienced rolling blackouts due to a shortage of electric power. At the same time, a potentially dangerous environmental

---

[2]Robert Malthus, *An Essay on the Principle of Population as It Affects the Future Improvement of Society* (New York: Random House, 1798, 1960), pp. 9, 13, 17.

threat was in the news—global warming caused by greenhouse gases released by the burning of carbon-based fuels.

Has Malthus finally been vindicated? Are we really running out of energy resources this time? Why did the energy crisis of the 1970s happen, and why did it end? Why did an energy crisis recur in 2001, and why just in California?

Energy is the stuff of life. With it, we can accomplish practically anything; without it, we can do nothing. What happens if we run out? Should we use less now to make the fuel we have last longer? Where can we find more? And even if we find more fuel, what will happen to our climate if we burn it?

Like most other useful things, energy can be misused. Improperly handled, it may be enormously destructive. Whether energy is used for good or ill depends entirely on the knowledge and wisdom of those wielding it. It is vital, therefore, that everyone understands as much about the subject as possible. When it comes to energy, knowledge really *is* power.

WHAT IS ENERGY?

The best definition comes from the science of *physics*. Energy is the capacity to do work. Work is defined as force multiplied by the distance through which it acts.[3] In the United States, work is expressed in units of *foot-pounds*, while in Europe, work is expressed in terms of *newton-meters* (or *joules*). A joule is the amount of work done by a force of one *newton*[4] acting through a distance of one *meter*.

Power is the rate at which work is done, and is calculated by dividing work by the time taken to do the work. The faster the work is done, therefore, the more power is expended. In the United States, power is expressed in units of *foot-pounds per second* and in Europe as *joules per second* (or *watts*). Most Americans are familiar with the term *horsepower*, which is defined as 550 foot-pounds per second.

It is useful to think of energy in terms of work because the whole reason we want to control energy is for the work it can do for us. In fact, the word *energy* comes from the Greek words *en* meaning *in* or *at*, and *ergon* meaning *work*.

Energy is measured in the same units as is work—that is, foot-pounds or joules. However, in the United States, the most common unit of measure is the *British Thermal Unit*, or BTU. A BTU is defined as the amount of energy needed to raise the temperature of one pound of water by one degree Fahrenheit.

One BTU is equal to 778 foot-pounds and to 1,055 joules (see *Appendix* E— *Units* for more conversion factors).

---

[3]Pushing on an immovable rock may seem like hard work, but to the physicist, no work is done until the rock actually moves.

[4]A *newton* is defined as the force needed to accelerate a one kilogram mass at the rate of one meter per second per second.

The following chart shows the orders of magnitude of energy use as measured in BTUs:[5]

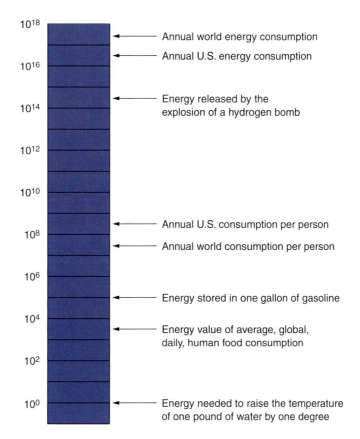

Key:
$10^0 = 1$
$10^1 = 10$
$10^2 = 100$
$10^3 = 1000$
...
$10^{18} = 1,000,000,000,000,000,000$

Notice that this chart does not use a linear scale. That is, each mark along the scale does not represent the same amount. Instead, each mark increases in value by a multiple (or *factor*) of 10. This is called a *logarithmic scale* and is useful for representing very large numbers.

To show the range between 1 BTU and 10^{18} BTU on a linear scale graph, we would need a piece of paper much longer than the distance between the earth and the sun!

[5]Adapted from Pennsylvania State University's Earth and Mineral Sciences website: www.ems.psu.edu/~radovic/fundamentals2.html

Huge amounts of energy are often measured in *quads*.[6] A quad is one quadrillion BTUs (that is 1,000,000,000,000,000; or 10^{15} BTUs). In 2001, the world used about 400 quads of energy, and the United States used almost a quarter of that amount.[7]

There are other useful ways in which to measure energy. The *kilocalorie* (or one thousand calories) is used to gauge the amount of energy contained in food. Typically, however, the word "Calorie" (with a capital "C") is commonly used instead of kilocalorie (abbreviated "kcal") when referring to food.

One thousand Calories (or one million "small c" calories), the amount of energy you might consume in a large meal, is equal to about 3,968 BTUs. One loaf of whole-wheat bread contains 1,440 Cal., or 5,714 BTUs, while one gallon of gasoline contains 126,000 BTUs. A person would have to eat 22 loaves of bread in order to be able to perform the same amount of work as a car engine running on a single gallon of gasoline.[8]

Another common energy unit is the *kilowatt-hour* used to measure electricity. One-kilowatt-hour is equivalent to 3,413 BTUs. Note that *kilowatts* measure capacity or flow, while *kilowatt-hours* measure quantity; a one-kilowatt electrical generator running for one hour produces one kilowatt-hour of electricity. It takes about one kilowatt of generating capacity to provide electricity to an average American home. The table on the following page shows the level of electrical energy consumption for uses ranging from residential to industrial.

Energy exists in two basic forms:

1. Potential—Energy at rest, waiting to be used
2. Kinetic—Energy in motion

A boulder sitting on the edge of a cliff is said to have *potential energy* by virtue of its position in the Earth's gravitational field. The boulder's potential energy is converted into *kinetic energy* when it is pushed over the edge, and gravity causes it to fall.

---

[6]This is especially true when comparing different fuels that may, separately, be measured in tons, barrels, or thousands of cubic feet.

[7]Although it appears from the chart that the U.S. accounts for nearly all of the world's energy consumption, remember that the scale is logarithmic, and for each unit up the scale, the value is increased by a factor of ten.

[8]Actually, an internal combustion engine converts energy into work more efficiently than does the human body. Taking the relative efficiencies into account, a person would have to eat almost 31 loaves to be able to perform the same amount of work that an engine can using one gallon of gasoline.

SCALES OF SELECTED ELECTRICITY USE, UNITED STATES

Use	Approximate Scale (kilowatts)
Portable radio	.0001
Cellular phone	.001
Portable computer	.01
Desktop computer	.1
Home (average)	1-1.5
Commercial customer (average)	10
Supermarket	100
Medium-sized office building	1,000
Medium-to-large factory	1,000-10,000
Largest buildings (peak use)	100,000
Largest industries (peak use)	1,000,000-10,000,000

Adapted from Seth Dunn; *Micropower: The Next Electrical Era*; Worldwatch Paper 151; (Washington: Worldwatch Institute, 2000), p. 32, Table 4.

Other objects that have potential energy are a gallon of gasoline (chemical energy), a stretched rubber band, a compressed spring, and a charged electrical battery. Things having kinetic energy include a moving car (matter in motion), a discharging battery (electrons in motion), sound (air molecules in motion), and a hot stove (atoms in motion).

WHERE DOES ENERGY COME FROM?

Most scientists think that it all started with something they call "The Big Bang." According to this theory, all the matter and space that now make up our universe were once compressed into a tiny, though incredibly dense, speck. Somewhere between ten and twenty billion years ago, the speck exploded—the Big Bang. With this tremendous explosion, the universe started an expansion that continues even now. Radiation filled expanding space as did hydrogen and helium gas.

Because the explosion was not perfectly uniform, the gases were not evenly spread out. Around a billion years after the Big Bang, some of the denser areas of gas started to condense into huge clouds. Gravity—the attractive force between matter—caused these clouds to collapse into vast spinning galaxies. Within these galaxies, smaller concentrations of atoms formed. As the density of these concentrations rose, heat was generated inside them. Temperatures increased until thermonuclear reactions were triggered, and stars were born.

Through a process known as *fusion*, stars combine the light elements, hydrogen and helium, into heavier elements such as oxygen, carbon, and iron. These elements are released into the universe when, in Carl Sagan's words, stars end "their lives in brilliant supernova explosions."[9] All of the elements that make up our planet and our own bodies were created in this way. We are, quite literally, made of stardust.

Our own star, the sun, is the source of most of the Earth's energy. The sun's energy comes to us as heat and light, which, in turn, give rise to other forms of energy. For example, the sun's uneven heating of the Earth's atmosphere is one cause of wind.[10] Heated air expands and rises and is replaced by cooler air in a process called *circulation*. Circulation produces wind as is illustrated below:

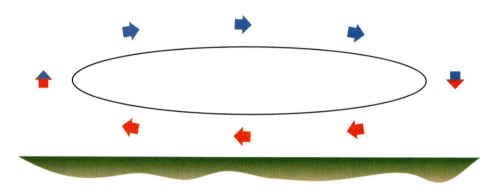

The ground absorbs the sun's heat, and warms the air above it. The warm air then rises and the cycle begins.

The sun's heat also causes water to evaporate. Water vapor condenses to form clouds and eventually falls back down to the Earth's surface as rain. Rain falling on high elevations and flowing downhill has kinetic energy that can be used to turn water wheels and turbines.

Sunlight provides the energy that plants need to live. Plants use a process called *photosynthesis* to turn sunlight, water, and carbon dioxide into food and oxygen. Nearly all animals live off of plants, either by eating them directly, or by eating other animals that eat them. So the sun indirectly provides the energy that powers our own bodies.

Sunlight also helped create the fuel that heats our homes and powers our machines. Wood and other plant material provided people with ready fuel for

---

[9]This quotation, along with the description of the Big Bang, comes from Carl Sagan, *Cosmos* (New York: Random House, 1980), pp. 246–47.

[10]The rotation of the Earth around its axis is another.

hundreds of thousands of years. In addition, all or most of the carbon-based fuels buried in the Earth's crust are probably a result of the sun's light. According to this theory, petroleum and natural gas were formed over millions of years in the oceans that cover most of the Earth. When the tiny plants living in the prehistoric oceans (and the tiny animals that lived off of them) died, their remains drifted down to the muddy ocean floor. There, bacteria caused them to decay. Tiny particles of sand and mud, called sediment, covered the plant and animal matter. As the sediment piled up, the remains were placed under enormous pressure that, in turn, generated heat. Most scientists believe that this combination of bacterial action, heat, and pressure transformed the plant and animal remains into crude oil and natural gas.

Coal was formed in a similar fashion when thick layers of dead plants piled up in swamps and rotted, turning into a substance called *peat* (which is itself used as a fuel in many parts of the world). When layers of sediment covered the peat, the resulting pressure transformed it into coal.

Because peat, coal, tar, bitumen, petroleum, and natural gas are believed to come from long dead plants and animals, they are often called *fossil fuels*.

A few scientists believe that at least some oil and gas may have been formed through nonbiological means. The chief American proponent of this theory is Thomas Gold, the astrophysicist who correctly predicted the existence of pulsars.[11]

At first, most scientists dismissed the idea out of hand, but new evidence has convinced some that nonbiological hydrocarbons do exist.

The theory may explain a number of mysteries, including the existence of hydrocarbons at extreme depths, in nonsedimentary rock, and on other planets.[12] It might also explain why some petroleum reservoirs seem to be refilling. Estimated Middle Eastern reserves, for example, have more than doubled in the last 20 years despite five decades of intense production and few additional discoveries.

If Gold's theory is true, then oil and gas could exist in the Earth's crust in vast amounts and could serve as a primary energy source for millennia.

Nonsolar sources of the Earth's energy include:

- *Tidal energy* caused by the sun's and the moon's gravitational pull.
- *Geothermal energy* produced by radioactive decay deep within the Earth and transmitted to the surface by the hot, molten layer of rock beneath the Earth's crust.
- *Fission*, the *atomic energy* released by the splitting of the nuclei (or cores) of uranium or plutonium atoms.

---

[11]Pulsars are rotating neutron stars that emit beams of light and radio waves. These waves are detected as pulses as they sweep over the Earth.

[12]Thomas Gold, *The Deep Hot Biosphere* (New York: Copernicus, 1999).

Corbis

- *Fusion*, the atomic energy released when atoms combine, as occurs in the sun's core.
- *Chemical energy* from *exothermic reactions*. For example, heat is released when potassium and water come into contact.
- *Cosmic radiation* (also called *background radiation*) left over from the Big Bang and from distant stars. The amount of this type of energy that reaches Earth is very small.

A BRIEF HISTORY OF ENERGY

According to Greek mythology, the Titans were giant, immortal beings. Two of the Titans, Prometheus (whose name literally means *forethought*) and his brother Epimetheus (*afterthought*), were tasked with giving different powers to the animals to help them survive. Snakes received venomous fangs, bears enormous strength, and deer great speed. But when man's turn came, no gifts were left. Moved by the helplessness of primitive people, Prometheus stole fire from the gods and gave it to the humans. Zeus, king of the gods, was so angered that he chained the Titan to a mountain, where he remained for thousands of years until Hercules freed him.

Underlying this story is the realization that, unlike animals, man is not well-equipped to adapt to nature. To survive, we must adapt nature to ourselves.[13]

---

[13]While humans are, perhaps, the only creatures that consciously *adapt* nature to fit their needs, virtually all living things inadvertently *alter* the environment by virtue of what they eat and emit, and by what eats them. It's a good thing for us that they do. If plants had not changed the atmosphere by producing oxygen, animal life as we know it would be impossible.

> ❝*Reasonable men adapt themselves to their environment; unreasonable men try to adapt their environment to themselves. Thus, all progress is the result of the efforts of unreasonable men.*❞
>
> **George Bernard Shaw, Irish playwright and writer**

With the story of Prometheus, the Greeks expressed the immense importance of fire in their lives. Indeed, the ancients saw fire as the spark of life and listed it among the four basic elements (earth, fire, water, and air) that they thought made up the universe.

Before fire, people's only source of power came from their own muscles. Fire brought warmth and light to cold, dark nights. It gave protection against animals far swifter and stronger than any man. The earliest evidence of fire's use was discovered in China, and dates back about 500,000 years.

Hundreds of thousands of years would pass before the next big leap—the domestication of animals. Evidence from China and southwestern Asia suggests that dogs were tamed there about 12,000 years ago. Sheep, goats, and pigs were domesticated around 8,000 B.C.; cattle in 6,000 B.C.; and horses, donkeys, and water buffalo in 4,000 B.C.[14]

By harnessing the power of the larger animals, people became far more productive. Oxen could plow land deeper and more quickly than a man or woman pushing a stick. Suddenly, more land could be cultivated than ever before and more crops grown on each acre.

With a more adequate and secure food supply, people began settling in one place. This enabled them to create and accumulate new, better, and larger tools (nomadic peoples were limited to what they could carry on their own or their animals' backs). As a result, advances in energy technology began to appear more quickly. Even so, with the possible exceptions of the sail, the windmill, the waterwheel, and gunpowder, the technology used by the average person did not change much for thousands of years. Romans living at the time of Christ would have easily understood the science of the 16th Century.

It was not until the late 17th Century, with the invention of the steam engine, that technology took off. After that, as the timeline in *Appendix* A shows, the history of energy (and, with it, people's lives) started changing at a furious pace.

---

[14]Jared Diamond, *Guns, Germs, and Steel: The Fates of Human Societies* (New York: W. W. Norton, 1997), p. 167.

ANIMATE ENERGY

Humans and animals were the first control-lable sources of work. A horse or ox could do the work of several people, but (by a grim logic) in places where water and arable land were scarce, slaves were preferred because they converted food into work more efficiently than did draft animals.[15] *Source:* U.S. Department of Agriculture.

PATTERNS IN HISTORY

Appendix A reveals a number of important patterns. The first is that the history of energy is really the history of our material development. People build machines to harness energy and magnify their ability to do useful work. This magnification has so increased human productivity that, in the West at least, children no longer must work in order to eat,[16] the elderly can look forward to a secure retirement, and women have been placed on an equal footing with men.[17] As Julian Simon wrote, energy truly is the *master resource.*[18]

Over time, people advanced to more efficient machines and to more efficient, concentrated, portable, and convenient forms of energy—from human muscles, to burning wood, to animal power, to wind and water power. Then, from these sources, humanity moved to whale oil and coal; next to petroleum, natural gas, and nuclear energy. Each advance left people better off and further from the hand-to-mouth existence that had been their lot for hundreds of thousands of years.

---

[15]J. R. McNeill, *Something New Under the Sun: An Environmental History of the Twentieth Century World* (New York: W. W. Norton, 2000), pp. 11–12.

[16]Child labor did not begin with the Industrial Revolution. In the centuries before the machine age, one person using a stick for a plow simply could not produce enough food to support a family. Consequently, everyone in the family had to work or face starvation. By increasing the productive power of an individual, the Industrial Revolution created the conditions under which child labor could be outlawed.

[17]Technology makes brains more important than brawn.

[18]Julian Simon, *The Ultimate Resource* (Princeton, NJ: Princeton University Press, 1981), p. 91.

The timeline in *Appendix* A also illustrates the fact that far more discoveries and inventions have occurred in the past two hundred years than in all of the hundreds of thousands of years that went before. Much of this is because as one discovery or invention leads to another, innovation piles upon innovation ever more rapidly.

Yet the great outpouring of creativity during the last two hundred years also coincided with the growth of personal and economic liberty in the world. This should not be surprising because individuals who are free to act and to enjoy the fruits of their actions have a strong incentive to invent things to increase their productivity and wealth. Slaves have no such incentive; creativity dies without freedom.

Most of the advances made in the last two hundred years were made in western countries. It is no coincidence that these countries were also the freest nations in the world during those two centuries. The right of individuals to own and trade property at mutually agreeable prices is essential for the efficient allocation of resources necessary for technological progress and economic growth.

> **❝**It is not surprising that new products almost always come from the free world and that communist countries, which contain one third of the world's population, account for only 3 percent of technological innovations.**❞**[19]
>
> **Mark Skousen, 1991**

The last and most important insight arising from the timeline is that people are very inventive. The appendix highlights only a few of the more important discoveries and inventions related to energy; a complete list would fill volumes. Throughout history, people have been faced with difficult problems, and throughout history, they have found solutions. Often, these solutions left them much better off than they were before the problems appeared. Strength is forged in adversity.

While a timeline can be useful for putting history into perspective and for revealing historical patterns, it can also be misleading. Timelines make history appear as if it were a logical progression, with one advancement leading inevitably to another—"The March of Progress." Yet history is not predetermined; ideas drive history, and not history ideas. Reality is messy, and progress, when it occurs at all, moves in fits and starts. Often great discoveries are ignored, and sometimes inventions are made, forgotten, and then invented again. In 60 A.D.,

---

[19]Mark Skousen, *Economics on Trial: Lies, Myths, and Realities* (New York: Irwin, 1991), p. 228.

Hero, a scientist living in Egypt, described the first known steam engine. But no practical use was made of this knowledge until steam engines were reinvented in the late 17th Century.

> ❝*To know and not to do is the same as not to know.*❞

Chinese proverb

But the most misleading thing about the timeline is that it suggests that an invention or discovery just appeared on a certain date in history as if by magic. For example, we learn that the Wright brothers flew the first powered airplane on December 17, 1903. But this bare fact conveys nothing of the years of toil, failure, and ridicule that the brothers endured. Nor does it tell us of the thousands of dreamers and inventors who went before; who—through their knowledge, creativity, sweat, and sometimes even their deaths—helped make that first flight possible. In the words of Sir Isaac Newton, "If I have seen further, it is by standing upon the shoulders of giants."[20]

---

[20]Letter to Robert Hooke, February 5, 1675 or 1676.

USING ENERGY

2

Putting Energy to Work

As explained in the previous chapter, work is force multiplied by the distance through which it acts. As this definition suggests, the trick in getting energy to do work is to channel it in such a way that it moves something. A simple example is a ship's sail. To get to the other side of a lake, one can either row a boat or put up a sail and let the wind do the work.

A windmill works in much the same way as does a sail, but instead of producing linear movement, it converts the wind's kinetic energy into rotary motion. Windmills have been used for centuries to perform such tasks as pumping water and grinding grain. One problem with a windmill is that it is an "intermittent resource"—it only works when the wind is blowing.

Waterwheels are similar to windmills, though their task is to turn the energy of falling or flowing water into rotary motion. While somewhat more reliable than windmills, early waterwheels had problems of their own. A shop, factory, or mill usually could not depend upon its waterwheel more than 160 days a year because of ice, floods, droughts, and dams that silted up.[21]

Besides their unreliability, another big problem with both windmills and waterwheels is that they are stationary and cannot be used to directly power a vehicle.[22] Enter the steam engine. Water can be boiled to generate high-pressure steam anytime and anywhere. The steam's energy can then be directly converted into rotary motion (with a steam turbine), or can be used to push a piston back and forth. The heat required to produce steam can be

---

[21]David Freeman Hawke, *Nuts and Bolts of the Past: A History of American Technology*, 1776–1860 (New York: Harper & Row, 1988), p. 195.

[22]They can, however, power vehicles indirectly. Windmills and waterwheels (or, at least, water turbines) can run generators to charge batteries that, in turn, power vehicles.

ELECTRICITY
--the modern
Emancipator

PUBLIC SERVICE COMPANY
OF NORTHERN ILLINOIS

generated by burning wood, alcohol, or carbon-based fuels (i.e., peat, coal,
oil, or natural gas); or with controlled *nuclear reactions*.

Internal combustion engines burn fuel directly inside piston cylinders where
expanding combustion gases drive the pistons.

Rotary motion created by these various means can be used to run machinery (looms, presses, and drills), turn wheels (cars, trucks, locomotives),
drive propellers (boats and airplanes), or to generate electricity.

In 1821, Michael Faraday discovered that moving a magnet near a coil of
wire produced a flow of electrons—an *electric current*—within the wire. This is the
principle of the *electric generator*: wrap a large coil of wire around a rotor, place the
rotor between some strong magnets, turn the rotor, and—presto—electricity!

ELECTRICITY

Electricity is an extremely versatile, portable, and convenient form of power, and about a third of America's primary energy is used to generate it. This proportion is expected to grow as computers and other information-age products continue to expand into more and more areas of our lives. The following sections describe conventional and alternative methods of producing electricity and the pros and cons of each.

GROWING HOME USE OF ELECTRICITY

This brief timeline of electricity's home uses shows how new applications of this versatile form of energy have continually emerged to increase efficiency and convenience and to make our lives both more productive and rewarding.

1900s	1910s	1920s	1930s	1940s
• Heater	• Refrigerator	• Air conditioning	• Electric razor	• Electric blanket
• Washing machine	• Electric trains	• Radio	• Can opener	• Dehumidifier
• Vacuum cleaner	• Hair dryer	• Blender	• Garbage disposal	• Electric guitar
• Cloths iron	• Christmas lights			
• Toaster				

1950s	1960s	1970s	1980s	1990s
• Television	• Jacuzzi	• Personal computer	• Dustbuster	• Internet
• TV remote control	• Self-cleaning oven	• VCR	• Rechargeable batteries	• Digital answering
• Dishwasher	• Microwave oven	• Waterbed	• Halogen torchierelamp	• Sony Play Station
	• Security system	• Crockpot	• Cellular telephone	• DVD player
		• Fax machine	• Noise machine	
		• Laser printer		

ENERGY USAGE: FROM LUXURY TO NECESSITY

Growing affluence has spread the use of automobiles and major electric appliances to most American households. *Source:* Compiled from *It's Getting Better All the Time: 100 Greatest Trends of the Last 100 Years,* by Stephen Moore and Julian Simon. Copyright 2000 by the Cato Institute. Reprinted by permission.

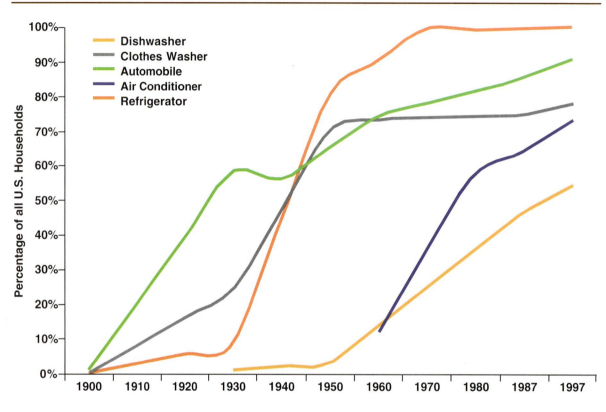

U.S. ELECTRICITY GENERATION—2002

While coal is still the fuel most often used to generate electricity in the United States, natural gas's market share is growing. Nine percent of United States power is generated by renewables, and only a fourth of that comes from nonhydro sources such as wind, solar, geothermal, and biomass. *Source:* U.S. Energy Information Administration, *Annual Energy Review 2002* (Washington: Department of Energy, 2003), p. 224.

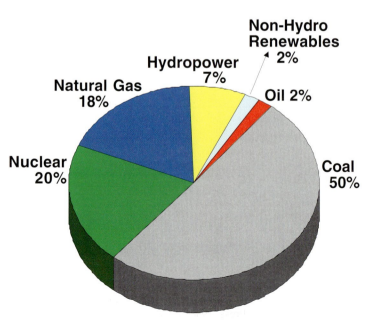

COAL-FIRED PLANTS[23]

The following diagram illustrates the basic operation of a conventional steam plant. Such plants generate most of America's electricity, and most are fueled by coal. In fact, more than half of the country's power comes from coal, the most plentiful carbon-based fuel.

In a typical plant, powdered coal is burned to boil water, converting it into high-pressure, superheated steam. The steam enters a turbine where it expands and drives the turbine's blades. The blades turn a shaft connected to a generator that creates electrical current.

After the steam leaves the turbine, it passes through condensers that cool it back into liquid water. The water is then pumped back into the boiler to repeat the cycle.

Coal comes in four basic forms: *anthracite, bituminous, subbituminous,* and *lignite,* as shown in the following table.

STEAM TURBINE DIAGRAM

This diagram shows how a steam power plant works. Nuclear plants work in much the same way—the only difference is the way in which water is heated to produce steam. Courtesy of *Power* magazine.

Steam Turbine Diagram

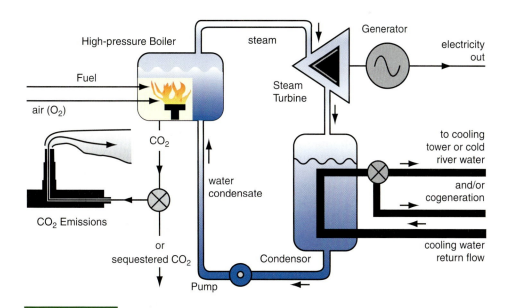

---

[23]Most of the information for this section on coal comes from Edward Cassedy and Peter Grossman, *Introduction to Energy: Resources, Technology, and Society* (Cambridge, UK: Cambridge University Press, 1998), chapter 6.

Form of Coal	Energy (BTU/lb)	Typical Sulfur Content (%)	Est. U.S. Reserves (billions of tons)
Anthracite	12,500	0.6	7
Bituminous	11,500	2.2	240
Subbituminous	9,500	0.5	180
Lignite	7,000	0.7	40

Adapted from Edward Cassedy and Peter Grossman, *Introduction to Energy: Resources, Technology, and Society,* p. 138, Tables 6.1 and 6.2.

Coal is essentially carbon plus some hydrocarbons and a minor amount of minerals. When coal is burned, heat is produced from the carbon and ash from the minerals. The higher the carbon content, the more heat and less ash that are produced when the coal is burned.

Anthracite has the highest BTU content, but it is also the least plentiful form of coal. Bituminous coal is the most common; its geographically widespread reserves make it readily available to power plants around the country. Unfortunately, it also has the highest sulfur content and releases more sulfur dioxide when burned.

Subbituminous coal has a relatively high BTU content and low sulfur, but 90 percent of its reserves are located in Montana and Wyoming, adding significantly to the cost of transporting it to power plants near major population centers.

Because of its relatively high sulfur content, bituminous coal lost market share to subbituminous coal after passage of the 1990 amendments to the Clean Air Act. Between 1990 and 2002, bituminous production dropped 18 percent (to 566 million short tons from 693 million tons), while subbituminous production rose 82 percent (to 445 million short tons from 244 million).[24]

Lignite, sometimes known as "brown coal," has a lower heating value than subbituminous coal and burns less cleanly. Consequently, it is not as desirable a fuel as are the other types of coal.

Coal use has had a larger effect on the environment than either oil or natural gas, though its impact has decreased with improving technology and stricter regulations. Coal affects the environment in four ways:

1. Extraction   Surface and subsurface mining can significantly alter the landscape and pollute groundwater. In strip mining, the *overburden* (the dirt and rock covering shallow coal seams) must be removed before the coal can be extracted. While under-

[24]U.S. Energy Information Administration, *Annual Energy Review* 2002 (Washington: Department of Energy, 2003), p. 203.

Corbis

ground mining results in less damage to the landscape, surface *subsidence* can occur as the tunnels collapse once the coal is removed.

Water seeping into abandoned mines may react with chemicals in the remaining coal to form acids that leach into underground aquifers and drain into rivers and lakes.

These problems can be handled through land reclamation (replacing the topsoil scraped off during strip mining operations) and back-filling empty tunnels to control subsidence.

2. Transportation

The nation's cleanest coal is located in the sparsely-populated West and may need to be moved long distances before it can be used. Railroads transport most of America's coal, but trucks and barges are also commonly used.

Coal also can be transported by pipeline. The coal is ground into powder, then mixed with water to form a slurry that can be pumped. However, this method requires a lot of water, which may not be available near a given mine.

3. Combustion

Burning coal produces pollutants including:

- Sulfur dioxide (SO_2)
- Oxides of nitrogen (NO_x), where x is an integer denoting the particular compound)

- Particulate matter (PM), usually called "particulates"
- Carbon monoxide (CO)

Coal also produces more carbon dioxide (CO_2) per BTU of electricity generated than do other fossil fuels.[25]

Emissions can be controlled in a number of ways. Electrostatic precipitators and filters ("bag houses") remove particulates; "scrubbers" eliminate sulfur dioxide and some nitrogen oxides.

Carbon dioxide removal technologies are still in their infancy, leaving power plant efficiency improvement as the most effective method of reducing CO_2 emissions.

4. Waste Disposal

Unburned ash must be removed from coal-fired plants and dumped. In addition, scrubbers produce large amounts of sludge that present disposal problems.

Currently, most coal ash is sent to landfills. Some is used to backfill mine tunnels, but this is only economical in cases in which the power plant is near the mine. The electric industry is looking for ways to put coal ash to productive use (for instance, the ash may be mined for sulfur and trace metals).

Controlling coal's impact on the environment is expensive. Ultimately, the costs are passed on to consumers in the form of higher prices for both coal and the electricity produced from it. However, as environmentalists point out, these higher prices better reflect the real societal costs of using coal. In the jargon of economics, *negative externalities* (unpaid costs) are being *internalized* (borne by the user). The prices also shift the burden of reducing coal's environmental impact to those who benefit from it.

Despite the higher prices resulting from environmental controls, coal is competitive with other fuels as a primary source of electric power (in 2002, coal-fired plants produced 50 percent of the electricity in the United States).

---

[25]While carbon dioxide is not a pollutant, it has been associated with global warming, a subject discussed in chapter 6.

COAL-FIRED POWER PLANTS
This chart shows the locations and capacities of the coal-fired plants in the U.S. Most plants are in the East near large population centers. However, most low-sulfur coal deposits are located in the West. *Source:* Bob Schwieger and Melissa Leonard, "First Annual Top Plants Survey," *Power,* August 2002, p. 62. Courtesy of *Power* magazine.

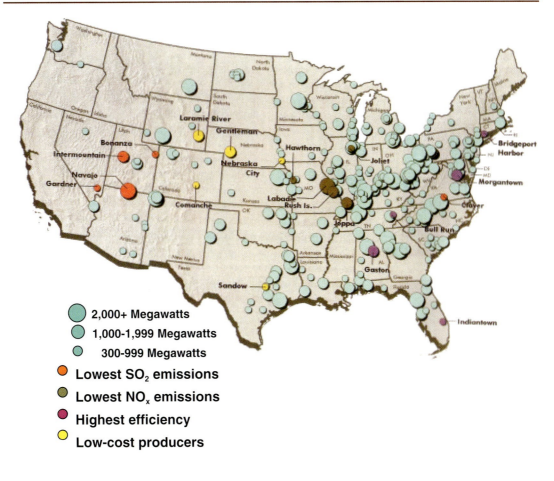

- 2,000+ Megawatts
- 1,000-1,999 Megawatts
- 300-999 Megawatts
- Lowest SO_2 emissions
- Lowest NO_x emissions
- Highest efficiency
- Low-cost producers

NUCLEAR FISSION

Nuclear power plants produce electricity in much the same way as do traditional power plants. Water is heated to produce steam to drive a turbine that, in turn, spins a generator. The big difference lies in how nuclear plants create the steam.

During *fission*, the nuclei of heavy atoms, typically uranium (^{235}U), are split into lighter nuclear parts. Energy is released in the process. Neutrons from one splitting atom strike other atoms, causing them to split in turn. Because

Corbis

more than one neutron is released when an atom is split, the fission reaction becomes multiplied in a *chain reaction*. The amount of fissionable material (like uranium or plutonium) needed to create a self-sustaining chain reaction is called a *critical mass*.[26]

Control rods of neutron-absorbing cadmium or graphite are used to regulate the chain reaction. By inserting or withdrawing the rods, the reaction is either slowed or increased.

The 104 active nuclear power plants in the United States produce about 20 percent of the country's electricity. There are another 350 nuclear plants throughout the rest of the world. Altogether, these plants produce about 18 percent of the world's electric power.

In the United States, the amount of electricity produced by nuclear plants has increased by 25 percent during the 1990s even though the number of nuclear plants fell by eight (from 112 to 104) during this same period.[27] This was made possible by raising the average *capacity utilization factor* of the remaining plants to 89 percent from 69 percent.[28] Put another way, the amount of time that the units were running versus their theoretical maximum rose by one-third.

Part of this improvement was due to better technology and techniques. But another reason was a change in *incentives*. Under traditional public utility regulation, the rewards that plant-owning utilities received were not tied to their performance. Firms simply received from customers, in addition to their

[26]An amount of mass that is more than that required to achieve a chain reaction that is *exactly* self-sustaining is called "super-critical." Anything less is called "sub-critical."

[27]U.S. Energy Information Administration, *Annual Energy Review* 2002, p. 257.

[28]Ibid., p. 224.

operating costs, an allowed rate of return based on the plant's "book value" (original cost minus depreciation). Under deregulation, nuclear plants now earn more money when they produce more power, so better performance means higher returns to shareholders.

In 1979, an accident occurred at the Three Mile Island (TMI) nuclear plant near Harrisburg, Pennsylvania. Although no one was injured and no harmful levels of radioactive emissions were released, the operating utility nearly went bankrupt paying for the cleanup. After the incident, public opinion turned solidly against the nuclear power industry. Nuclear power was dealt an additional blow in 1986 when a terrible accident at the Chernobyl plant in the Soviet Union killed 31 people immediately and exposed an estimated 4,000 more to high doses of radiation.

Containment vessels built around American reactors are designed to be an ultimate safeguard against any incidents. The vessel at TMI worked as designed; but tragically, Chernobyl, built under far lower safety standards than are the norm in western countries, had no such last line of defense.

Advances in technology offer the possibility that future reactors can be made inherently safe from meltdown, but existing reactors of older design will remain in operation for many years. While the U.S. industry has taken steps to reduce the possibility of human error, some analysts argue that accidents due to operator mistakes are inevitable.

Potentially, the biggest problem with nuclear power is the management and disposal of the tons of radioactive wastes produced every year. Nuclear plants produce far less waste than do coal plants. A 1,000-MW nuclear-electric plant, for example, produces about one metric ton of waste per year, versus one *million* tons from a similarly sized coal plant. However, nuclear waste is far more dangerous. Many of the waste products are highly toxic and remain radioactive anywhere from less than one year to millions of years. On the other hand, toxicity is generally inversely proportional to half-life, and some scientists argue that after about 1,000 years most of the waste would be no more dangerous than uranium ore.

Further complicating the storage problem is that the wastes initially generate large amounts of heat. Spent fuel currently is stored at the plants in pools of water that absorb the radiation and dissipate the heat. Heat production drops quickly as the wastes age.

Geological isolation is the only viable long-term disposal solution currently available. This means storing the wastes in highly stable geologic formations that have remained seismically inactive for millions of years. Such formations exist both on land and beneath the oceans, but transporting spent fuel to these sites must be done with care. Terrorists could try to sabotage storage sites or, more likely, attack convoys hauling the materials.

The so-called NIMBY (Not In My Back Yard) syndrome is just as important as are the geological issues in locating a suitable site for waste disposal. Not surprisingly, people are reluctant to live near a nuclear waste dump. Billions of dollars have gone into building a permanent storage facility for high-level radioactive wastes at Yucca Mountain, Nevada. The facility, run by the Department of Energy, was to have opened in 1998, but the project is behind schedule. Even after construction is complete, however, political opposition from Nevada's citizens and politicians may keep the facility's doors shut.

For years, nuclear engineers have argued that spent fuel should be reprocessed to extract any unconsumed uranium along with the plutonium, neptunium, and lawrencium created during fission. These extracted elements could then be fed back into power plants as fuel. While *pyroprocessing* may be technically feasible, current reactors are not capable of using the low-grade fuel thus created.

NUCLEAR POWER PLANTS

This chart shows the locations and sizes of the nation's nuclear power plants. Most of the facilities are located east of the Mississippi River and near large population centers. *Source:* Bob Schwieger and Melissa Leonard, "First Annual Top Plants Survey," *Power,* August 2002, p. 44. Courtesy of *Power* magazine.

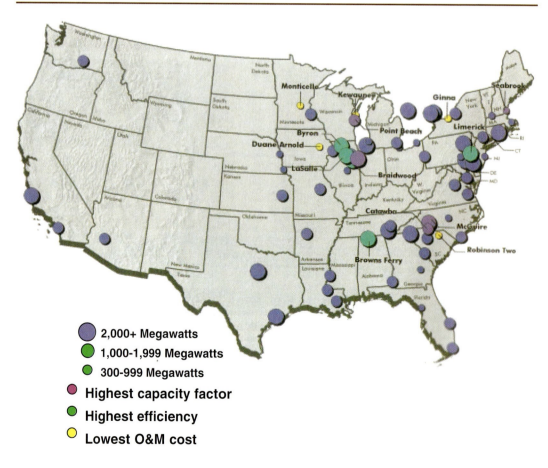

- 🔵 **2,000+ Megawatts**
- 🟢 **1,000–1,999 Megawatts**
- ⚪ **300-999 Megawatts**
- 🔴 **Highest capacity factor**
- 🟢 **Highest efficiency**
- 🟡 **Lowest O&M cost**

Nuclear power plants are much more expensive to build than conventional plants, but their operating and maintenance (O&M) costs are less. It is possible, however, that nuclear power would not be viable without the type of government support that began in the 1950s and 1960s. This support has taken a number of forms, including:

- Direct subsidies. Beginning in 1957, the U.S. Atomic Energy Commission (now the U.S. Nuclear Regulatory Commission) helped pay some of the construction costs of plants built by private utility companies.
- Research and development. Since the establishment of the U.S. Department of Energy in 1977, the government has spent more than $20 billion on nuclear power research and development. In fact, the first commercial reactor was based on reactors designed for use in U.S. Navy submarines.
- Accident liability limits granted to power companies under the Price-Anderson Act of 1957, as amended.

NATURAL GAS PLANTS

Natural gas is the cleanest of the fossil fuels. It leaves no residue and produces less pollution than either oil or coal.

Natural gas is used in both gas turbine and steam generating plants. The most efficient way to use it is in a *combined-cycle* system. In such plants, fuel is burned in a combustion chamber to produce hot, high-pressure gases that pass directly through a gas turbine that, in turn, powers a generator. The still-hot gases are then sent to a waste heat boiler where they heat water to produce steam. The steam turns a turbine that is connected to a second generator. Spent steam is piped to a condenser where it is cooled back into water. The water is pumped back into the boiler, repeating the cycle.

Natural gas became the fuel of choice for new electric generation in the 1980s and 1990s due to falling gas prices and significant efficiencies in gas-fired combined-cycle technologies. As recently as 2001, energy forecasters were predicting that the market share of gas would double in the next two decades.[29]

However, the prices paid for natural gas by power generators have increased by over 50 percent since 2000, while coal prices have dropped. The Energy Information Administration now forecasts that coal growth in the U.S. will be nearly equal to that of gas through 2025.[30]

---

[29]National Energy Policy Development Group, *National Energy Policy* (Washington: Government Printing Office, 2001), p. 1–7.

[30]U.S. Energy Information Administration, *Annual Energy Outlook* 2004 (Washington: Department of Energy, 2004), pp. 82–83, 135.

NATURAL GAS COMBINED-CYCLE POWER PLANTS

Gas-fired combined cycle power plants, the newest plants in the country, are located in the nation's fastest growing areas. *Source:* Bob Schwieger and Melissa Leonard, "First Annual Top Plants Survey," *Power,* August 2002, p. 38. Courtesy of *Power* magazine.

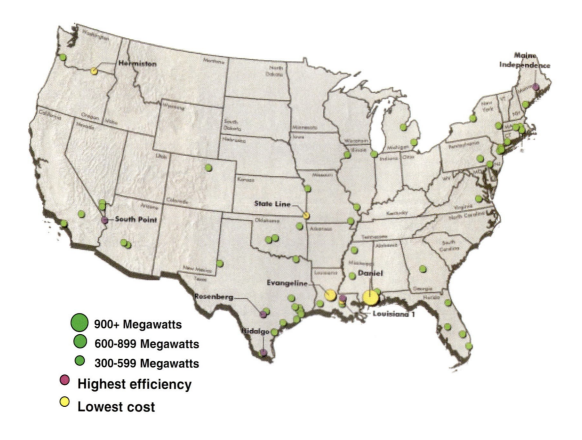

- ● 900+ Megawatts
- ● 600-899 Megawatts
- ● 300-599 Megawatts
- ● Highest efficiency
- ○ Lowest cost

HYDROELECTRIC PLANTS

Hydroelectric plants produce electricity by releasing falling water through turbines that drive generators. Despite the fact that hydroelectric plants produce no pollution, they have fallen out of favor with many in the environmental community who point out that dams disrupt local ecology, place large tracts of land (often including wildlife habitat) under water, and interfere with the migration of indigenous fish.

Currently, hydroelectric plants produce about 7 percent of both America's and the world's electricity. The following table shows estimates of hydroelectric generating capacities that could technically be exploited versus those al-

Corbis

ready being used. These numbers are somewhat misleading because only about half of this capacity would be economical to use. Presumably, most (if not all) of the portion that has already been exploited comes from the economic half.

VIABLE HYDROPOWER SITES—POTENTIAL AND EXPLOITED

Region	Technically Exploitable Potential (gigawatts)	Already Exploited (gigawatts)	% Used
Asia	610	98	16
South Asia		45	
China		33	
Japan		20	
Latin America	432	96	22
South America		85	
Central America		11	
Africa	358	17	5
North America	356	148	42
Canada		58	
United States		90	
Former USSR	250	62	25
Europe	163	145	89
Eastern Europe		17	
Western Europe		128	
Oceania	45	12	27
World Total	2,214	577	25

Source: Edward Cassedy, *Prospects for Sustainable Energy: A Critical Assessment* (Cambridge, UK: Cambridge University Press, 2000), p. 137.

OIL-FIRED PLANTS

Most oil-fired plants work in much the same way as coal-fired steam plants, although petroleum (like natural gas) can also be used to power turbine generators.

While 39 percent of America's overall energy came from oil in 2002, less than 3 percent of the country's electricity was generated from oil-fired plants.[31]

Oil resources are less plentiful and generally more expensive than coal, but oil has a lower environmental impact. It burns more completely than coal and leaves no ash to be hauled away. It also produces fewer emissions per unit

---

OIL/GAS STEAM POWER PLANTS

Steam plants that can be fueled by either oil or gas are older and less efficient than the newer combined-cycle facilities. Most are used only during high-demand peak periods as swing capacity, and most use gas rather than oil in order to meet air quality regulations. *Source:* Bob Schwieger and Melissa Leonard, "First Annual Top Plants Survey," *Power,* August 2002, p. 54. Courtesy of *Power* magazine.

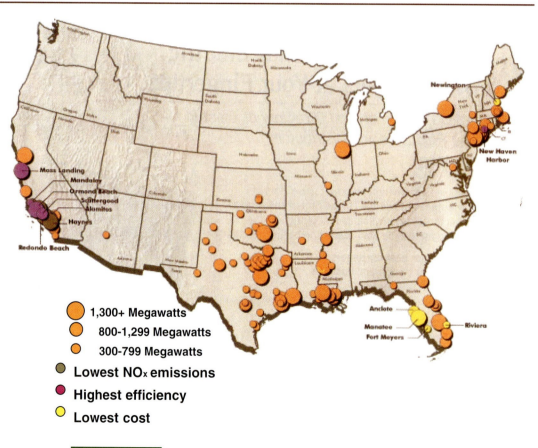

- ● 1,300+ Megawatts
- ● 800-1,299 Megawatts
- ● 300-799 Megawatts
- ● **Lowest NOₓ emissions**
- ● **Highest efficiency**
- ● **Lowest cost**

---

[31]U.S. Energy Information Administration, *Annual Energy Review* 2002, pp. 9, 225.

of energy generated. Oil wells leave a much smaller footprint than do coal mines, and advances in directional drilling have significantly reduced this footprint even more.[32]

Oil is cheaper to transport than coal because it can be more easily pumped through pipelines.

WIND POWER

Wind power is favored by many environmentalists as the best alternative to power generation from carbon-based fuels. Modern wind turbines use blades, modeled after airplane propellers, to turn electric generators.

Power output is proportional to the cube of the wind speed, and directly proportional to the area swept by the turbine's blades. A typical unit has a blade diameter of about 33 meters (108 feet) and has a capacity of 400 kilowatts. However, General Electric (GE) has developed 3.6-MW turbines designed for offshore use.

Installation costs run about $1,000 per rated kilowatt, not counting transmission lines. This cost and the turbines' operating availability of about 95 percent compare favorably with conventional power plants. However, because turbines work only when the wind is blowing, annual production under even the best conditions is generally only about 20 percent to 35 percent of rated capacity.[33] Adjusting for these numbers, the installed cost of a turbine is closer to $3,000 to $5,000 per kilowatt.

Denmark is the world's leader in wind-power technology; nearly 15 percent of that country's electricity comes from wind turbines. In addition, Denmark supplies state-of-the-art turbines to countries around the world including the United States. Backing from the Danish government has, in part, accounted for the prominence of wind power there. In February 2002, however, the government announced that it was ending its subsidies due to the high cost.

Several areas in the United States boast wind conditions suitable for the operation of *wind farms*, including Iowa, Kansas, North Dakota, and South Dakota. The consistent winds in California's Altamont Pass near San Francisco have made it the country's leading site. California took the lead in wind power in this country with an aggressive tax credit program during the early 1980s. While

---

[32]Directional drilling techniques allow a drill bit to be guided from the wellhead to oil reservoirs not directly below the wellhead. These techniques are especially useful for offshore production and production in environmentally sensitive areas because they enable multiple well bores to fan out in many directions from a single platform.

[33]Chris Namovicz, *Update to the* NEMS *Wind Model*, presentation to the NEMS/AEO 2003 Conference in Washington, D.C., on March 18, 2003. See www.eia.doe.gov/oiaf/aeo/conf/pdf/namovicz.pdf, slide 15.

some early "tax farmers" used the tax breaks to cash in on failed wind farms, by 1993 wind power supplied more than one percent of the state's electricity.[34]

Problems with wind power include:

- *Location*. Areas with favorable wind conditions are not always near demand centers, and the value of the power produced by remote wind farms may not be worth the cost of building transmission lines.
- *Unreliability*. The intermittent nature of the wind makes turbines unsuitable as primary power sources, at least until significant advances in storage technology are made.
- *Land use*. For wind farms to produce significant amounts of energy, they must incorporate hundreds of turbines requiring large tracts of land (though each turbine's small footprint does allow the land to be used for agriculture or grazing). Assuming that the wind blew all the time, it would take *twenty-five hundred* 400-kW wind turbines to replace one traditional 1,000-MW power plant. Given that the wind does not blow all the time, however, and assuming a capacity factor of 33 percent, it would take 7,500 turbines to replace a traditional plant! By a similar calculation, it would take 835 of GE's 3.6-MW turbines installed offshore to replace a regular facility.

WIND POWER: THEN AND NOW

Machines that convert wind energy into electricity are not new. At the left is an illustration of an experimental wind station from an 1891 issue of *Scientific American*. To its right, a picture of a standard model from the late 20th century. Courtesy of Enron Corp.

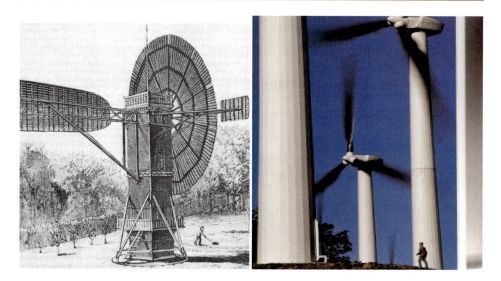

---

[34]Christopher Flavin and Nicholas Lenssen, *Power Surge: Guide to the Coming Energy Revolution* (New York: W. W. Norton, 1994), p. 119.

- *Aesthetics*. Some people object to wind farms because they block their view of nature. Sunlight filtering through rotating turbine blades can also produce an irritating stroboscopic effect.
- *Harm to wildlife*. Birds are killed when they fly into the rotors. The kills, while relatively small in number, may be significant for endangered species such as golden eagles. Of particular concern have been such prime wind areas as Altamont Pass, California and Tarifa, Spain.
- *Noise* (not a factor if the turbines are installed in remote areas).
- *Safety*. Wind turbines can throw ice that builds up on the blades, and the blades themselves can come loose. Blade and turbine maintenance often must be performed at dangerous heights.

At least some of the objections to wind power, such as land use, aesthetics, and noise, might be overcome by placing windmills offshore. However, a proposal to site 170 turbines five miles off the coast of Massachusetts was attacked on the basis that it "would 'industrialize' the area, interfere with local fishing, destroy a 'place of pristine relaxation' for boaters and drive away tourists," according to an article in *The Wall Street Journal*.[35] While the United Kingdom and other coastal European countries have moved ahead with offshore projects, the United States has yet to launch one of its own.

GEOTHERMAL ENERGY

The Earth's core is a vast and essentially unlimited source of heat. Most of this heat is at depths that are currently beyond our reach, but some exists near the surface. Water in these zones can be extremely hot (up to 2,200°F) and under

PhotoDisc

[35]John Fialka, "Florida Utility Finds It's Not Easy Even Trying to Be Green," *The Wall Street Journal*, April 4, 2002, p. A20.

very high pressure. If wells are drilled into these formations, the water can be brought to the surface and used to drive turbines. If water does not exist naturally in such hot formations, it can be pumped in through injection wells and then back out after it has been heated.

The steam and water produced in this manner often contain salts and minerals that pit and corrode turbine blades. Equipment operating under these conditions is subject to frequent breakdowns and high maintenance costs.

Along with salts, the water from geothermal wells, commonly called brine, may contain toxic elements such as lead, arsenic, boron, mercury, and gases such as hydrogen sulfide, which is extremely toxic. The brine may be handled by re-injecting it into the ground. Water re-injection would also eliminate any land subsidence that might otherwise occur.

One problem with geothermal energy is its limited availability. There are few areas on Earth suitable for geothermal power generation. In the United States, geothermal power production is centered in a few western states, and plants are often located in environmentally sensitive areas such as national parks. Another problem is that geothermal sites tend to cool down with use.

Country	Geothermal Generating Capacity in 1998 (megawatts)
USA	2,850
Philippines	1,850
Italy	770
Mexico	740
Indonesia	590
Japan	530
New Zealand	350
Iceland	140
Costa Rica	120
El Salvador	110
Nicaragua	70
Kenya	40
China	30
Turkey	20
Portugal (Azores)	10
Russia	10
Other	10
Total	**8,240**

Source: Ernest McFarland, "Geothermal Energy," in John Zumerchik, ed., *Macmillan Encyclopedia of Energy,* 3 vols. (New York: Macmillan Reference USA, 2001), vol. 2, Table 1, p. 576.

MICROTURBINES

Microturbines are small combustion turbines about the size of a refrigerator. They can produce anywhere from 25 to 500 kilowatts—enough to power 25 to 500 homes. Although they are typically fueled by natural gas, the turbines can also run on diesel. The use of microturbines and other remote devices is known as *distributed generation*.

Microturbines fill an important niche by providing power to areas far away from existing power grids. In developing countries especially, micro power can help bridge the technology gap between the old and new worlds. However, some experts like Johannes Pfeifenberger believe that, "[e]ven with the continued technological progress of distributed resources,. . . the base-load and intermediate-load markets most likely [will] remain dominated by central station power plants."[36] This is because of:

- Economies of scale in development costs (siting, permits, and fuel contract negotiations are less expensive when handled in quantity). Moreover, it takes far fewer resources to build a single 1000-MW generator than two thousand 500-kW units.
- Economies of scale in control, operations, and maintenance.
- Load balancing (i.e., by pooling many customers, the chances of coincidental peak loading is reduced).
- Modular designs that improve plant operating flexibility and allow them to grow with demand.
- Decreasing cost of pollution controls.[37]

In the late 1800s, during the early years of the American electrical power industry, *isolated plants*—"systems designed to light a single building and operated from a 'powerhouse' in the basement"[38]—dominated the market. Such small stations were fast, easy, and cheap to build, and provided quick returns on investment.

Thomas Edison, however, was convinced that the future lay with *central power stations*. He realized that an isolated powerhouse would have to be built with enough capacity to handle its building's peak load. This meant that most of the system's capacity was unneeded for much of the time. Edison reasoned that by pooling users and connecting them to a centralized generating station, the peaks and valleys of demand would even out, and the capacity of the single station could be far less than the sum of the total capacities of the individual powerhouses.

---

[36]Johannes Pfeifenberger, "What's in the Cards for Distributed Resources?," *The Energy Journal*, *Distributed Resources Special Issue*, International Association for Energy Economics, 1997, p. 15.

[37]Ibid., p. 5.

[38]Forrest McDonald, *Insull* (Chicago: University of Chicago Press, 1962), p. 26.

BENEFITS OF MICROPOWER

Benefit	Description
Modularity	By adding or removing units, micropower system size can be adjusted to match demand.
Short lead time	Small-scale power can be planned, sited, and built more quickly than larger systems, reducing the risks of overshooting demand, longer construction periods, and technological obsolescence.
Reliability and Resilience	Small plants are unlikely to all fail at once. They have shorter outages, are easier to repair, and are more geographically dispersed. Factories and computer facilities often use microturbines for backup in case of a loss of power from the regular service provider.
Avoided plant and grid construction, and losses	Microturbines can be placed at the site where power is needed, thus eliminating the need to construct expensive power transmission lines (although the need for a fuel distribution system to supply the microturbines remains). Local siting also eliminates grid power losses.
Avoided emissions and other environmental impacts	Small-scale power generally emits lower amounts of particulates, sulfur dioxide and nitrogen oxides, and heavy metals, and has a lower cumulative environmental impact on land and on water supply and quality.

Source: Adapted from Seth Dunn, *Micropower: The Next Electrical Era,* Worldwatch Paper 151, Washington, July 2000, p. 33, Table 5.

Around 1910, Samuel Insull, the father of the modern power generation industry and a protégé of Thomas Edison, ran a series of studies on the use of isolated powerhouses located in Illinois. As his biographer, Forrest McDonald explained, Insull found that, "For all Illinois outside of Cook County, the combined demand for power was just over 300,000 kilowatts, and various users had installed about 437,000 kilowatts to cover it, though they never used more than about 225,000 kilowatts at the same time. If they were connected as an integrated system, they could easily be served with a total capacity, including abundant reserve, of about 270,000 kilowatts, thus saving more than half of the $43,000,000 that was invested in power supply in the area. The waste of fuel under the existing systemless arrangement was incalculable, but Insull believed that at least three to four times as much coal as necessary was being burned."[39]

The pattern of starting with remote power plants and gradually shifting to central power stations is being repeated in developing countries today.

Proponents of micropower argue that with today's technology, *distributed* does not have to mean *isolated.* Distributed power sources can be tied into the local grid. When home or business owners do not need their generators' total capacity, excess power can flow into the grid and be sold at a profit. Once

[39]Ibid. 142.

enough distributed power supplies are tied to a grid, the need for central plants could actually disappear. Under such a scenario, utilities would not sell power, but instead would sell access to the grid just as internet providers now sell access to the world-wide web.

SOLAR POWER

Photovoltaics, or solar cells, convert sunlight directly into electricity. When photons strike certain semiconductor materials, such as silicon, they dislodge electrons. These free electrons collect on the specially-treated front surface of the solar cell, creating a potential difference between it and the back surface. Wires attached to each of the cell's faces conduct the current. Individual cells can be combined in panels to increase voltage.

Because solar cells only work when the sun shines, they must either be used together with storage devices or as supplements to conventional facilities.

Photovoltaics have provided energy for spacecraft and for remote devices such as floating buoys. However, because of their high cost, they are still not practical for large-scale power generation.

Total grid-connected solar power generation in the United States is currently only about 60 megawatts with perhaps another 120 megawatts not connected to the grid.[40] This totals to less than half the capacity of a traditional mid-sized plant. The few central solar generation facilities in operation are experimental and use large tracts of land. With current technology, about 100 square feet of photovoltaic (PV) panels are required to generate one kilowatt of electricity in bright sunlight. It would take hundreds of square miles of solar panels to replace an average nuclear power plant.

About 10,000 square miles (26,000 km^2) of PV panels would produce enough electricity to supply all U.S. electrical and non-electrical energy needs, while 85,000 square miles (220,000 km^2) would be needed to supply the world with power. By contrast, as an article in *Science* pointed out, "all the PV cells shipped from 1982 to 1998 would only cover [about] 3 km^2," or about 1.16 square miles.[41]

Some scientists suggest that the size of the solar power footprint could be reduced by as much as 75 percent by placing satellites in space to collect sunlight, convert it into electricity, and then beam the power to the Earth's surface in the form of microwaves.[42]

---

[40]U.S. Energy Information Administration, *Annual Energy Outlook* 2004 (Washington: Department of Energy, 2004), p. 156. Communication from EIA to authors, March 16, 2004.

[41]Martin Hoffert, et al, "Advanced Technology Paths to Global Climate Stability: Energy for a Greenhouse Planet," *Science*, November 2002, p. 984.

[42]Ibid.

Corbis

Currently, researchers are concentrating on two aspects of the solar cell technology: making solar cells less expensive, and making them more efficient. Unfortunately, high efficiency and low cost tend to be mutually exclusive.

Another way of harnessing solar power is to use an array of mirrors to concentrate, or focus, sunlight onto water flowing through a metal pipe. The resulting steam can then be used to drive a turbine.

BIOMASS

Biomass energy is derived from plants or animal wastes. Wood, a form of biomass, was the first fuel used by humans, long before coal or any of the hydrocarbons. Wood was used for heating caves and later homes, and, much later, for powering steam engines. Usually it was not used in a renewable manner, and many forests were cut down faster than they grew back. Vast oak forests in California's Central Valley, for example, were cut down to fuel early locomotives.[43]

Today, wood and other biomass is more often, but not always, a renewable energy source, and often the biomass used for fuel is the byproduct of other processes. Biomass power accounts for about two-thirds of the nonhydroelectric renewable energy generated in the United States, producing about 3 percent of the country's electrical power. Counting both electric and nonelectric usage, biomass accounts for approximately 4 percent of U.S. and 7 percent of the world's total primary energy use.

---

[43]Environmentalists have complained that most uses of biomass around the world "are neither renewable nor sustainable." Christopher Flavin and Nicholas Lenssen, *Power Surge*, p. 177.

PhotoDisc

Biomass is a very broad term that covers many primary sources, electric generation technologies, and alternative fuels for transportation. It includes crops grown specifically for energy purposes (so-called *energy crops*) and residue or waste materials, also called *opportunity crops*. Energy crops include fast-growing trees (e.g., eucalyptus) and corn (used for ethanol production). Opportunity crops are more varied and include lumber mill waste, paper mill residues, food crop residues (both field harvesting residue and food processing residue), forest thinnings (e.g., to reduce fire risk), and animal wastes.

Currently in the United States, nearly all biomass electric power generation—probably 90 percent to 95 percent—is based on wood type fuels. Examples include wood waste from pulp, paper, and lumber mills. Examples of biomass not derived from wood include agricultural wastes such as sugar cane bagasse, rice and nut hulls, and fruit pits.

Biomass fuels can be either burned directly to produce steam to drive electric generators, or first converted to a solid, liquid, or gas fuel. Conversion may be by thermal, chemical, or biological processes, or some combination of these methods. Biological processes, like fermentation, covert biomass materials into fuel forms, such as natural gas or gasoline substitutes. Thermal processes like gasification decompose the biomass into combustible gaseous fuels similar to natural gas.

Electric power generation and heating are the main uses of biomass in the United States. A small amount of biomass is converted to ethanol fuel for transportation. Beyond electrical power generation, biomass fuels are used for industrial, commercial and residential heat. The primary wood products industries dominate in the use of biomass heat, with pulp and paper applications surpassing sawmill and lumber applications within the sector. Since the 1970s, the pulp and paper industry has increasingly used leftover materials as fuel to generate steam and power for the paper making process.

During the 1980s, many wood-fired, and a smaller number of municipal-waste-fired, electric power plants were constructed. In several cases, electric utilities built wood-fired power plants or converted existing coal-fired power plants to burn wood and mixtures of wood and coal.

Cofiring—mixing biomass with coal—is the most economical, near-term technology for biomass, with a potential of approximately 7,000 MWe in the United States.[44] The potential economic benefits of cofiring include reduced coal consumption, reduced SO_2, and NO_x emissions, and additional revenue received from wood waste disposal. Negative impacts include increased costs, reduced efficiencies, and potential lost power due to the lower heat density of the biomass. Some biomass fuels also produce increased emissions of hydrogen chloride and heavy metals such as lead, cadmium, and mercury. Reduced marketability of the resulting fly ash may also be a factor.

Another constraint on biomass cofiring is emerging as more coal-fired plants are required to adopt selective catalytic reduction (SCR) as a NO_x control technology. The catalysts used in SCR may be poisoned when exposed to alkali-containing flue gas from biomass.

Plants and trees are able to turn only about 1 percent to 3 percent of the sun's energy into usable fuel, and only a fraction of that can be turned into work by burning them. Even less energy would be produced if the plant matter were first converted into methanol or bio-diesel and then burned. Considering that solar cells' energy efficiency ranges from 15 percent to 20 percent, it is clear that more energy can be produced by covering the ground with photovoltaics than with trees.[45]

There is little reason, therefore, to grow crops specifically for the purpose of energy production, although cultivation of dedicated energy crops is increasing rapidly in some areas due to heavy government subsidies. On the other hand, using residue biomass that would otherwise be wasted does make economic sense, and, in fact, opportunity fuels by far account for most of the energy use of biomass in the developed world.

While biomass is the oldest fuel known to man, the technologies to grow and harvest biomass continue to improve and mature. In addition to the gasification of biomass to produce natural gas and transportation fuels, even more advanced systems, similar to petroleum refineries, can produce a wide variety of products simultaneously, including electricity, plastics and pharmaceuticals, and heat. These "biorefineries" can also produce useful products from the ash and can use a wide variety of biomass materials as input.

---

[44]MWe and MWt are used to measure plant capacity. MWe indicates megawatts of electrical output, and MWt megawatts of thermal output.

[45]Bjørn Lomborg, *The Skeptical Environmentalist: Measuring the Real State of the World* (Cambridge, UK: Cambridge University Press, 2001), p. 134.

TIDAL POWER

Electric power can be generated from the water flow caused by rising and falling tides. Only a few experimental tidal plants exist in the world today, although a number of suitable locations have been identified. In general, such plants cost significantly more to build than do conventional facilities, and they provide only intermittent service (i.e., when the tide is either coming in or going out).

Tidal power has been used for centuries. As far back as 1734, a mill in Chelsea, Massachusetts used four tide-driven waterwheels to grind spices. It is estimated that under optimum conditions the installation may have generated as much as 50 horsepower.[46] As with wind power, however, tidal power could not survive the introduction of inexpensive electricity generated from carbon-based fuels.[47]

FUEL CELLS

Fuel cells work on the same principles as do storage batteries, except that free electrons are provided by the continuous flow of some fuel-like hydrogen rather than by the corrosion of an electrode.

Fuel cells are efficient and clean; their only effluent is pure water.[48] They have few moving parts and are therefore quiet, reliable, and maintenance free. Like microturbines, fuel cells have found a niche in providing distributed power for remote sites and in serving as power backups.

Also, because fuel cells are so reliable, some companies are using them to power critical computer systems. Momentary power surges and declines, which can cause significant computer problems, are common in today's grid systems. The trade journal, *Public Utilities Fortnightly*, reported that "In 1997, the First National Bank of Omaha switched from the grid to fuel cells after experiencing a costly computer crash at its data processing center."[49]

One of the main drawbacks in using fuel cells is their high cost, which is about $5,200 per kilowatt of capacity as compared to $1,300 to $1,500 per kilowatt for a diesel generator.[50] Another drawback is the difficulty in supplying the hydrogen fuel that powers all fuel cells.

---

[46]Wilson Clark, *Energy for Survival: The Alternative to Extinction* (Garden City, NY: Anchor Books, 1974), p. 331.

[47]Ibid.

[48]If hydrogen fuel is provided by a reformer, a device that breaks a hydrocarbon fuel such as natural gas or gasoline into hydrogen and carbon dioxide, then carbon dioxide is also emitted.

[49]Ruth Kretshmer and Kenneth Hundrieser, "Reliability: What Level and What Price?," *Public Utilities Fortnightly*, November 1, 2001, p. 15.

[50]U.S. Energy Information Administration, *Assumptions to the Annual Energy Outlook*, February 2004, p. 35.

Despite these problems, fuel cells have generated interest among auto makers and even some oil companies. This attention brings with it additional focus on both wind and solar power. Some hope that these environmentally friendly or green power sources can provide the electricity required to extract the hydrogen that fuel cells need from either water or methane.

Some of the enthusiasm over fuel cells has begun to die down, however. Part of this may be due to the pall that fell over the high-tech sector when the dot-com bubble burst on Wall Street in 2001.

According to the International Energy Agency, "fuel cells are . . . projected to make a modest contribution to global energy supply after 2020, mostly in small decentralized power plants. . . Fuel cells in vehicles are expected to become economically attractive only towards the end of the projection period. As a result, they will power only a small fraction of the vehicle fleet in 2030."[51]

FUEL CELL TECHNOLOGIES

Type	Description
Alkaline Fuel Cell	Uses a potassium hydroxide electrolyte and operates at 400°F. Alkaline fuel cells have the highest electrical efficiency (70 percent), but are too costly for commercial use. These types of fuel cells are used aboard NASA space shuttles.
Proton Exchange Membrane Fuel Cell (PEM)	Uses a polymer membrane electrolyte and can generate anywhere from a few watts to hundreds of kilowatts. Their relatively low operating temperatures (about 200°F) make these fuel cells suitable for residential and automotive applications. PEM fuel cells include the Direct Methanol Fuel Cell that extracts its hydrogen fuel directly from methanol (eliminating the need for a reformer[52]).
Phosphoric Acid Fuel Cell	Uses a phosphoric acid electrolyte and operates at 400°F. This type of fuel cell was the first to be employed in commercial stationary power generation. Because of its flexibility, the cell is suitable for use by hotels, hospitals, airport terminals, and even locomotives and buses.
Molten Carbonate Fuel Cell	Uses a potassium/lithium carbonate electrolyte and operates at about 1,200°F. Molten carbonate fuel cells have electrical efficiencies of 50–55 percent. Suitable for megawatt-size applications such as commercial buildings and institutions.
Solid Oxide Fuel Cell	Uses a zirconium dioxide ceramic electrolyte that allows the highest operating temperature (1800°F) of all types of fuel cell (higher temperatures generally mean higher efficiencies).

Source: Adapted from Carl Levesque, "How Soon is Now? Looking for Fuel Cell Technology's Future," *Public Utilities Fortnightly,* Vol. 139, No. 20, November 1, 2001, p. 25.

---

[51]International Energy Agency, *World Energy Outlook*: 2002 (Paris: OECD/IEA, 2002), p. 30.

[52]A reformer is a device that extracts hydrogen (and carbon dioxide) from hydrocarbon fuels such as methane and gasoline.

NUCLEAR FUSION

Like fission, nuclear fusion converts some of the mass in an atom's nucleus into energy. While fission accomplishes this by splitting nuclei, fusion does it by joining two nuclei. The sun is essentially a huge nuclear-fusion reactor.

Despite decades of research, no one has yet been able to create a sustained fusion reaction in the laboratory, and scientists do not expect to be able to construct an operating demonstration plant before 2020. If such plants are possible, they offer the promise of a clean and nearly inexhaustible supply of energy.

COMPARATIVE GENERATION COSTS

Which energy technology generates electricity at the least cost? Answering this question is difficult because of constantly changing fuel, maintenance, and regulation costs and because of differences in government subsidies and

COMPARISON OF POWER GENERATION COSTS FOR NEW CAPACITY (CENTS PER KILOWATT HOUR)
Electricity generated from oil, natural gas, or coal is cheaper than that generated from renewable resources. Fossil-fuel plants are also more reliable and have more flexibility in size and location.[53]

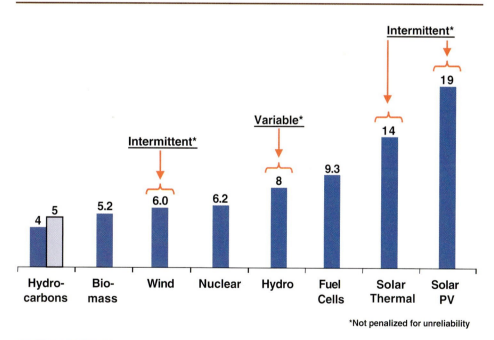

*Not penalized for unreliability

[53]U.S. Energy Information Administration, *Annual Energy Outlook* 2003 reference case aeo2003. d110502c. See Appendix F.

tax treatment. Many assumptions must be made, but the U.S. Energy Information Administration estimates that given current technology, hydrocarbon-fired generation is the cheapest and solar the most expensive.

However, availability is as important as the cost of production. The market places a higher value on technology that can reliably produce power at the instant it is needed. "Dispatchability" (industry jargon for the ability to deliver power on demand) is essential to satisfy customers. The availability of both wind and solar power varies from moment to moment, while hydropower, which is ultimately dependent on rainfall, varies by season and year. Thus the costs shown in the previous illustration understate the true costs of the less reliable technologies.

POWER TRANSMISSION

Electrical power is carried from generation plants over high-voltage wires. The use of high voltages reduces line losses over large distances. Before the power can be used at a home or business, its voltage must be reduced by a device known as a *transformer*.

Transmission lines are controlled by SCADA (supervisory control and data acquisition) systems consisting of remote sensors that transmit data about the lines and the power flowing through them to a central control station.

In the United States, transmission lines are interconnected to form *grids*. Linking the transmission lines together in this way allows power plants to back each other up in case of problems. As the map below illustrates, there are ten separate grids that supply power within the continental United States, Canada, and a portion of Baja California Norte, Mexico. With few exceptions, the grids themselves are not interconnected.

Corbis

Power transmission has long been considered to be a *natural monopoly*. The belief is that building two or more sets of competing transmission lines each capable of supplying power to every home or factory in a town would be wasteful and would result in higher prices to consumers. In order to avoid such duplication, government authorities typically grant monopoly rights to a single transmission company and then regulate the company's business decisions and the rates that it charges its customers.[54]

NORTH AMERICAN ELECTRIC RELIABILITY COUNCIL

A series of power outages, led by the northeast blackout of 1965, resulted in the 1968 formation of the North American Electric Reliability Council (NERC) and its ten Regional Reliability Councils (see Appendix F). Copyright 2004 by the North American Electric Reliability Council. Reprinted with permission.

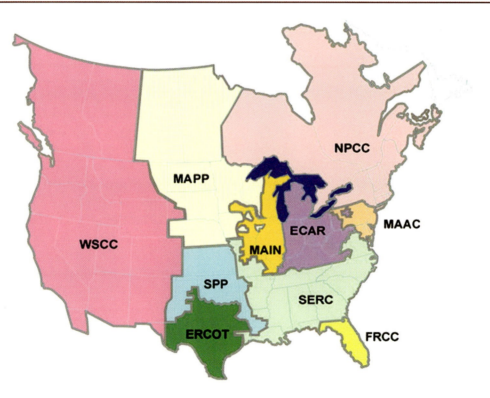

---

[54]Industry leaders a century or more ago successfully obtained what became known as the *regulatory covenant*—franchise protection from would-be competitors in return for maximum rates based on a cost-plus determination. Robert Bradley, Jr., "The Origins and Development of Electric Power Regulation," in Peter Grossman and Daniel Cole, eds., *The End of a Natural Monopoly: Deregulation and Competition in the Electric Power Industry* (New York: JAI, 2003), pp. 43–75.

Still, electric distribution companies directly competed in dozens of American cities as late as the 1960s.[55] Economists such as Walter Primeaux see benefits to open competition between so-called public utilities. Economists also recognize the imperfections of *public-utility regulation*, where one firm is given a legal monopoly and all but guaranteed cost recovery at varying levels of performance.[56]

TRANSPORTATION

Transportation accounts for more than a quarter of America's energy consumption and about a fifth of world energy use. The sections that follow discuss not only the primary power plant used in today's vehicles—the internal combustion engine—but also possible alternatives to the engine and to its most common fuels, gasoline and diesel.

INTERNAL COMBUSTION ENGINE

Much of the world's oil is used to move people and goods, and most of that is consumed by internal combustion engines. When the gasoline-fueled automobile first appeared, it was hailed as a great boon to the environment. That may seem strange today, but at the turn of the century horses and oxen powered most vehicles. Fueling a nation's draft animals requires that much land be placed under agriculture—resulting in a loss of natural habitat. In the early 1900s, "it took about 2 hectares [almost 5 acres] of land to feed a horse—as much as was needed by eight people. . . . In 1920, a quarter of American farmland was planted to oats, the energy source of horse-based transport."[57]

Worse, animal power turned city streets into filthy breeding grounds for disease, reeking of manure and urine and swarming with flies. San Francisco's ordinances still include a law that bans the piling of horse manure more than six feet high at street corners. Another legacy of horse power is the custom that a gentlemen walks to the outside when escorting a lady down a sidewalk. This was done to shield the lady's dress from any muck that might be thrown up by passing carriages.

---

[55]Walter Primeaux, Jr., "Total Deregulation of Electric Utilities: A Viable Policy Choice," in Robert Poole, ed., *Unnatural Monopolies: The Case for Deregulating Public Utilities* (Lexington, MA: Lexington Books, 1985), p. 128.

[56]The difficulties of regulation include determining when costs are *reasonable* and when new investments are *prudent* given the incentive of utilities to incur greater costs (such as salaries) and "pad" or "gold plate" the rate base (physical assets) to earn more profits.

[57]J. R. McNeill, *Something New Under the Sun*, p. 310.

This scene from a midwestern street around 1910 shows transportation in transition from the horse-and-buggy era to the age of the horseless carriage. The smoke of the early automobiles was considered much less polluting than the excrement and carcasses of horses on the street. *Source:* John Jakle and Keith Sculle, *The Gas Station in America* (Baltimore: John Hopkins, 1994), p. 207.

In addition to the tons of waste that had to be scraped off city streets and carted away each day, the bodies of thousands of dead horses had to be disposed of. "A big city had to clear 10,000 to 15,000 horse carcasses from the streets every year."[58] Early autos were noisy and belched smoke, but at least they kept the streets clean.

Today's engines are far more powerful, efficient, and cleaner than their ancestors. No other power plant can yet match the gasoline engine's combination of convenience, power, and low cost. Consider that a fifteen-gallon gas

Corbis

---

[58]Ibid.

tank gives the average car a range of more than 300 miles. When the tank is empty, it can be quickly filled at any one of tens of thousands of service stations across the country.

Despite occasional spikes, the price of gasoline has, on average, declined over the past 80 years (after adjusting for the decreasing purchasing power of the dollar due to monetary inflation). This decline is even more impressive when two key factors are considered. First, the quality of gasoline has improved greatly over the decades. Second, local, state, and federal retail motor fuels taxes have increased more than the rate of inflation.

In the United States, motorists pay an average tax of over $0.40 per gallon or 25–30 percent of the pump price. As high as this is, motorists in Canada pay more than $0.60 per gallon, motorists in Spain and Australia pay more than a dollar per gallon, and drivers in some European countries pay more than two dollars in taxes per gallon, giving them the most expensive motor fuel in the world.[59] Because Americans drive greater distances than do Europeans, and in larger

---

U.S. RETAIL GASOLINE PRICES 1917–2003 ($/GALLON IN 2003 DOLLARS)

Average U.S. gasoline prices have remained steady or declined over the decades despite motor fuel taxes that have risen more than the rate of inflation. *Source:* See Appendix F.

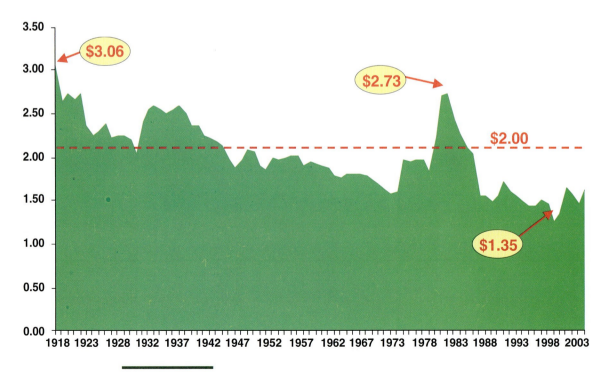

---

[59]International Energy Agency, *Energy Prices and Taxes, First Quarter* 2003. (Paris: OECD/IEA, 2003).

vehicles that use more fuel per mile traveled, it has been more difficult politically to raise taxes in the United States than it has on the other side of the Atlantic.

ELECTRIC CARS

Electric cars are not a new idea. In the late 1800s, most American automobiles in regular production were electric. In fact, according to historian David Kirsch, "the Electric Vehicle Company was both the largest vehicle manufacturer and the largest owner and operator of motor vehicles in the United States."[60] The cars were quiet, easy to operate, and could travel about 40 to 60 miles before needing to be recharged.

EARLY ELECTRIC VEHICLE ADVERTISEMENT

This advertisement highlights the attractions of early electric vehicles. *Source:* Advertising Ephemera Collection, Emergence of Advertising On-line Project.

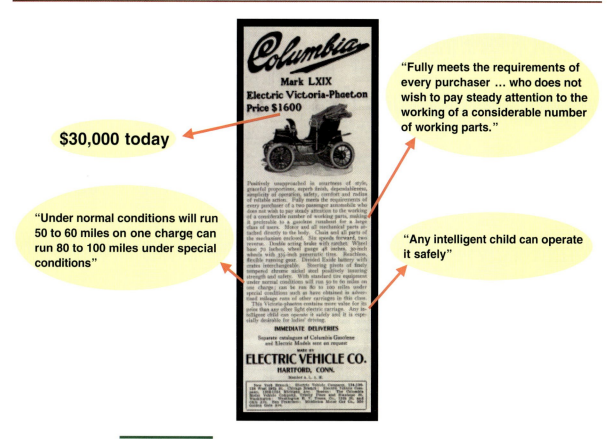

$30,000 today

"Fully meets the requirements of every purchaser ... who does not wish to pay steady attention to the working of a considerable number of working parts."

"Under normal conditions will run 50 to 60 miles on one charge, can run 80 to 100 miles under special conditions"

"Any intelligent child can operate it safely"

---

[60]David Kirsch, *The Electric Vehicle and the Burden of History* (New Brunswick, NJ: Rutgers University Press, 2000), p. 31.

Electrics soon fell out of favor with the driving public, however. Gasoline engines replaced electric motors as the power plant of choice because of their greater power and range. By 1914, internal combustion engines powered 99 percent of the 568,000 vehicles manufactured in America.[61]

While there have been advances in storage battery technology, batteries have not kept pace with the higher demands consumers place on cars.[62] Today's electrics (also called zero emissions vehicles, or ZEVs) can move faster than their early predecessors and offer more in the way of amenities, but their range is still limited to between 50 and 150 miles after five-hour battery charges.[63]

The vast majority of the auto trips most people make are within a few miles of their homes. Were range the only limitation, then, an electric would make an acceptable second or third car for those who can afford more than one vehicle. Unfortunately, ZEVs have other serious shortcomings:

- They can cost two or three times as much as comparable conventional vehicles, although ZEV proponents hope that mass production will eventually result in lower prices.

- In order to stretch their driving range, manufacturers have had to build ZEVs out of lightweight materials. Consequently, they do not stand up as well to collisions as do their heavier, gasoline-powered rivals.

- Typical ZEVs have load capacities of about half of those of conventional vehicles.

- The cars' batteries are expensive (about 20 percent of the cost of a vehicle) and must be replaced every four to six years.[64]

- The batteries contain toxic chemicals, including lead, and their disposal creates serious waste management problems.

- While the cars themselves do not pollute, the power plants that supply the electricity needed to recharge their batteries do. Some environmentalists have dubbed ZEVs "Emission Elsewhere Vehicles."[65]

Despite these limitations, in 1990 the California Air Resources Board's Low-Emission Vehicle Program mandated that by 2003 ten percent of all new cars sold in the state had to be electric. CARB relaxed the requirement

---

[61]Ibid. 15.

[62]The lead-acid battery is still the most cost-effective battery for cars even though it has been around for a century and a half.

[63]Edward Cassedy, *Prospects for Sustainable Energy*, p. 165.

[64]Ibid., p. 164.

[65]Amory Lovins, quoted in *Alternative Fuels: Myths and Strategies*, American Petroleum Institute, August 8, 1995, p. 3.

Electric Vehicles: Then and Now

Electric vehicles are still a niche product after a century of development. The similarity between early and some modern electrics underscores a lack of progress relative to vehicles powered by the internal combustion engine. *Source:* (a) Colorado Historical Society, Denver Art Museum. Denver Public Library. (b, c) illustration by Jean Spitzner.

(a)

(b)

(c)

several times, and, in January 2001, gave auto makers credits toward the 10 percent ZEV goal for partial zero emission vehicles (PZEVs), advanced technology zero emission vehicles (AT-ZEVs), and super-ultra-low-emission vehicles (SULEVs).[66]

Critics point out that these vehicles are so expensive that the only way that automakers can sell even this reduced number is to price them well

[66]U.S. Energy Information Administration, *Annual Energy Outlook* 2002 (Washington: Department of Energy, 2002), p. 17.

below cost. This means that the price of regular cars must go up to offset manufacturers' losses on the sale of electric vehicles. As a result, some consumers may be forced to drive their old cars longer than they ordinarily would. The net result could actually be dirtier air because older cars tend to be less efficient than new ones. In fact, it could well be that auto emissions could be reduced far more and with much less cost simply by helping owners of old, heavily-polluting cars replace their vehicles with newer, cleaner models.

Another possible unintended consequence of California's program may be more traffic deaths; older cars and ZEVs are not as safe as new conventional automobiles.

Despite such problems, California's Low-Emission Vehicle Program has been adopted by several other states, including New York and Massachusetts—states with dense traffic and associated air quality problems.

Even with these looming mandates, Ford Motor Company abandoned its two lines of Think! electric vehicles (see the above figure) in August 2002. After spending $123 million on development, the company concluded that there simply were not enough customers interested in the vehicles.[67]

HYBRIDS

Hybrid Electric Vehicles (HEVs) offer a viable alternative to all-electric cars. Hybrids are powered by an internal combustion engine and driven by one or more electric motors. Although a number of configurations are possible, typically the gas engine runs a generator that powers an electric motor at each of the car's wheels. An electric battery provides back-up power for entering traffic and passing, and is recharged by the generator when the vehicle is idling or operating at cruising speeds.

Although HEVs aren't true zero emission vehicles, they do offer a number of advantages over all-electric cars, including:

- Better acceleration
- Lower cost
- Greater range (500 to 700 miles)

---

[67]Micheline Maynard, "Ford Abandons Venture in Making Electric Cars," *New York Times*, August 31, 2002, p. B1.

- No need for lengthy battery recharges
- Fewer batteries to replace
- Fueled by readily available gasoline or diesel

HEVs also have some advantages over traditional cars:

- Lower emissions
- Significantly better gas mileage
- Greater range
- Similar or better performance

There are several reasons why hybrids are more efficient than conventional cars. First, their own internal combustion engines can be much smaller (and therefore lighter) because they need to be sized only for average operating conditions. Any additional needs are supplied by the battery.

Hybrids also recover some of the kinetic energy that is normally lost when braking. When traditional cars brake, they convert kinetic energy into heat, whereas hybrids use regenerative braking. During braking, the electric motors are switched to work as generators (a generator is essentially a motor working in reverse). The torque required to turn these generators is converted into electrical energy, which is fed back into the storage battery.

Finally, like ZEVs, HEVs rely on lightweight materials to reduce their overall weight.

In the near term, hybrids offer a much more realistic alternative to traditional cars than do all-electric vehicles.

ALTERNATIVE FUELS

LPG and CNG[68]

Liquefied Petroleum Gas (LPG) together with *Compressed Natural Gas* (CNG) are the most common alternatives to gasoline and diesel used in the United States. Both fuels produce fewer emissions than gasoline and about 25 percent less carbon dioxide (CO_2).[69] Until recent years, both fuels were less expensive than gasoline.

---

[68]CNG, the abbreviation for *compressed natural gas*, has been copyrighted by the company, Consolidated Natural Gas.

[69]T. Y. Chang, R. H. Hammerle, S. M. Japar, and I. T. Salmeen, "Alternative Transportation Fuels and Air Quality," *Environmental Science and Technology*, Vol. 25, no. 7 (1991), p. 1194.

LPG and CNG-fueled vehicles have found a number of market niches. Theme parks with heavy foot traffic sometimes use natural gas-powered carts to avoid the fumes that would be produced were the carts fueled with gasoline or diesel instead. Farmers have used LPG from on-site storage tanks rather than install gasoline pumps to refuel their vehicles. Fleets whose vehicles travel regular routes have also found natural gas to be an economical alternative to traditional fuels.

On the down side, the Department of Energy estimates that new natural gas vehicles (NGVs) can cost anywhere from $2,500 to $5,000 more than conventional vehicles, while LPG-fueled cars cost about $2,500 more.[70] In addition, these alternative-fuel vehicles (AFVs) are less reliable and less convenient to refuel. Finally, because both fuels contain less energy by volume than does gasoline, larger tanks are needed. In fact, a cubic foot of compressed natural gas contains only about one quarter of the BTUs that are in a cubic foot of gasoline. Tanks on an NGV can take up nearly all the available cargo space, while holding only about 150 miles worth of fuel.

Despite these drawbacks, the federal government and some states have promoted the use of AFVs through subsidies and tax breaks. In 2000, for example, Arizona began offering its citizens a lump-sum rebate of 40 percent of the price of a new AFV. Thousands of people took advantage of the program to purchase taxpayer-subsidized trucks. The program was abruptly ended seven months and $500 million dollars after it began.[71]

The environment may actually be worse off as a result because the program encouraged the purchase of trucks, which use more fuel than do regular cars. Worse, participants only needed to promise to use 100 gallons of alternative fuel a year in their vehicles in order to qualify, so many of the new trucks ended up burning mostly gasoline.

In Albuquerque, New Mexico, the police department purchased 15 natural gas-powered squad cars for $25,000 each with a grant from the federal government. As Sean Paige reported in *Reason* magazine, "The autos have only about half the range of conventional patrol cars, they perform sluggishly, and they can be refueled at only one location in town." Still, the department's fleet coordinator said, "We couldn't turn down what was basically a free car."[72]

---

[70]U.S. Department of Energy, *Taking an Alternative Route* (Washington: Department of Energy, 2001), pp. 18, 19.

[71]Sean Paige, "The Great Pickup Stick-Up," *Reason*, June 2001, p. 43.

[72]Ibid. 47.

ETHANOL

Ethanol is an alcohol produced from the fermentation of sugar. In the United States, it is typically made from corn. Benefits of ethanol over gasoline include:

- Lower carbon dioxide emissions (though other emissions are comparable)
- Non-toxic
- Renewable supply

Problems with ethanol include:

- About 20 percent less BTU content by volume.
- Cannot be transported through existing pipelines.
- Significantly more expensive to produce.
- Requires that a significant amount of land be placed under cultivation. Along with this would come an additional load on the fresh water supply, increased use of fertilizers (which could end up in streams and rivers), and loss of forestland and other natural habitat.
- Creating ethanol may consume more energy than is contained in the ethanol. This point is controversial, and probably cannot be resolved as long as the government subsidizes production of the fuel. If producing ethanol on the free market yields a net monetary profit, then it will likely yield a net energy profit as well (see the section, *Energy Economics* in the next chapter).
- Ethanol-fueled vehicles cost several hundred dollars more than comparable conventional vehicles.

"The fuel of choice during the early days of the automobile industry appears to have been none other than ethanol. . . . The availability of gasoline was very limited and its distribution system was not yet in place. Ethanol, by contrast, derived from the fermentation of sugars and starches, was a well established industry, and a relatively abundant supply of it was available. It is reported that in 1908, when Henry Ford began the production of the famous Model T, which was to establish the automobile as we now know it, he consulted with Thomas Edison whether to use gasoline or ethanol as the fuel for the new model vehicle. Edison advised Ford to choose gasoline.**"**[73]

John Ingersoll

---

[73]John Ingersoll, *Natural Gas Vehicles* (Lilburn, GA: Fairmont, 1996), p. 20.

EARLY ETHANOL ADVERTISEMENT

This advertisement for a 1902 French auto, cycle, and boat show, features alcohol as the fuel of choice. *Source:* Reproduced in John Ingersoll, *Natural Gas Vehicles* p. 21.

METHANOL

Methanol also is an alcohol, but, unlike ethanol, it is highly toxic. It can be made from coal, natural gas, wood, and biomass. Methanol's advantages over gasoline are:

- Lower carbon dioxide emissions (other emissions are comparable)
- Renewable supply

Its disadvantages include:

- Cannot be transported through existing pipelines
- Somewhat more expensive to produce
- More toxic
- About 50 percent less BTU content by volume, requiring larger fuel tanks and resulting in less vehicle cargo and passenger space
- More corrosive (harder on auto parts)
- Lower vehicle resale value
- More frequent vehicle oil changes
- Higher vehicle cost (several hundred dollars more)

In the 1970s and 1980s, methanol attracted government support as "the most promising alternative to motor vehicle fuel" for the United States.[74] The California Energy Commission promoted the fuel as a way to increase energy security by reducing dependence on petroleum imports while, at the same time, decreasing air emissions. Methanol's attraction faded in the 1990s as reformulated gasoline and improvements in vehicle technology significantly and affordably reduced emissions with no inconvenience to motorists.[75]

HYDROGEN

From an environmental standpoint, hydrogen is nearly an ideal fuel because its only products of combustion are water and some nitrogen oxides. Unfortunately, hydrogen is very reactive and does not exist in a pure state on Earth.

[74]*Methanol: Fuel of the Future, Hearing before the Subcommittee on Fossil and Synthetic Fuels*, 99th Cong, 1st sess., (Washington: Government Printing Office, 1986), pp. 43, 80, 114.

[75]Robert Bradley, Jr., "The Increasing Sustainability of Conventional Energy," *Cato Policy Analysis*, No. 341, April 22, 1999, p. 24.

Hydrogen is therefore considered to be an *energy carrier* (like a battery), rather than an *energy source*. Hydrogen cannot replace fossil fuels, nuclear power, or any other primary energy source. In fact, energy from these sources must be expended to produce hydrogen.

Hydrogen usually is extracted from hydrocarbons although it can also be generated by *water electrolysis*, a process that consumes a lot of electricity.

If the electricity used to produce hydrogen is generated by burning coal or hydrocarbons, there would be little environmental benefit over the use of re-formulated gasoline. Significant emission reductions would be achieved only if the electricity were to be produced by solar or wind power, a hydroelectric facility, or a nuclear plant.

There may prove to be a symbiotic relationship between hydrogen fuel use and both solar and wind power. One of the chief problems with these sources of electricity is that they are intermittent—they only work when the sun is shining or the wind is blowing. This makes them unsuitable as power sources for customers who require a steady supply of electricity. However, the sporadic nature of these sources is less of a problem for the purpose of hydrogen production.

Hydrogen can also be extracted from hydrocarbons such as methane or gasoline. Another promising method involves mixing borax with water to form sodium borohydride, and passing the mixture through a catalyst chamber to produce hydrogen.[76]

NASA scientist Friedemann Freund has suggested that there may be vast amounts of hydrogen existing in the top 12 miles of Earth's crust.[77] If this hydrogen can be economically extracted (i.e., if the hydrogen that is extracted contains more energy than must be expended to produce it), it may provide a nearly inexhaustible source of clean energy.

One problem with using hydrogen as a fuel is a phenomenon known as *hydrogen embrittlement*. Under high pressure and temperature, hydrogen, the smallest of the atoms, can flow into the intermolecular spaces in steel. When this occurs, the metal can become brittle and susceptible to fracture.

Also, as with other alternative fuels, there is no distribution network for hydrogen, so refilling the tank would present a problem.

---

[76]Julie Wakefield, "The Ultimate Clean Fuel," *Scientific American*, May 2002, p. 36.

[77]John Bluck, "Hydrogen-Fed Bacteria May Exist Beyond Earth," NASA *News*, April 3, 2002.

WELLS-TO-WHEELS COMPARISON

This diagram shows the flow of energy resources from the wells or mines to consumers, and illustrate the different infrastructures involved for conventional (current) and hydrogen technologies.

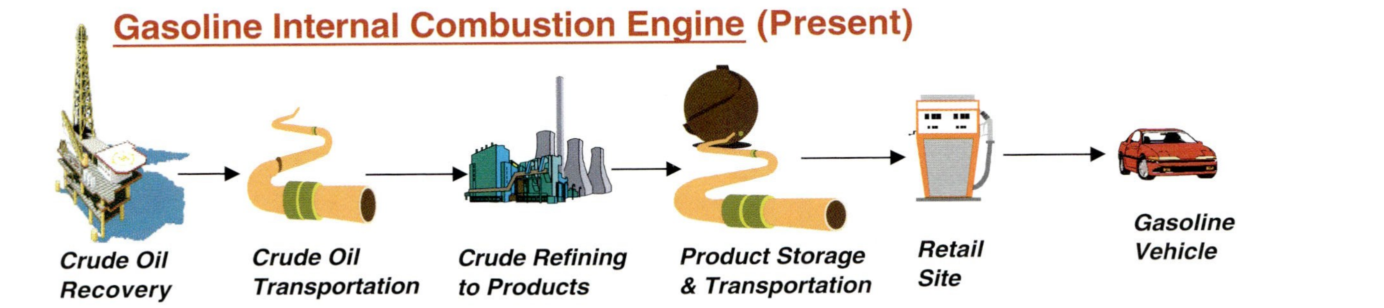

Gasoline Internal Combustion Engine (Present)
Crude Oil Recovery
Crude Oil Transportation
Crude Refining to Products
Product Storage & Transportation
Retail Site
Gasoline Vehicle

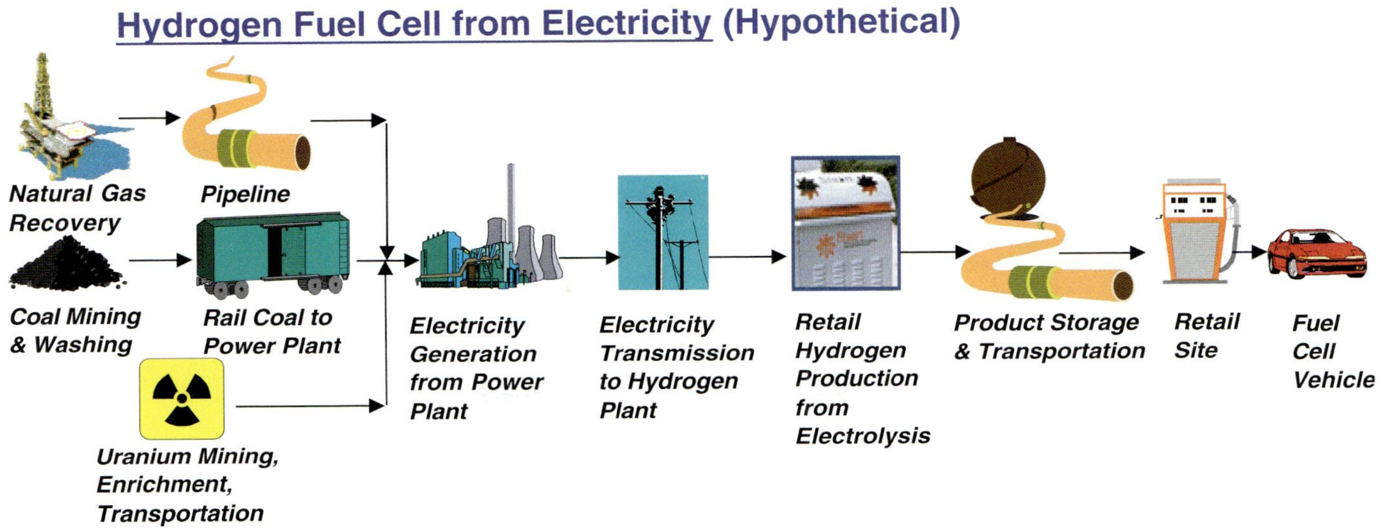

Hydrogen Fuel Cell from Electricity (Hypothetical)
Natural Gas Recovery
Pipeline
Coal Mining & Washing
Rail Coal to Power Plant
Uranium Mining, Enrichment, Transportation
Electricity Generation from Power Plant
Electricity Transmission to Hydrogen Plant
Retail Hydrogen Production from Electrolysis
Product Storage & Transportation
Retail Site
Fuel Cell Vehicle

EFFICIENCY— TECHNICAL AND ECONOMIC

3

FIRST AND SECOND LAWS OF THERMODYNAMICS

Energy cannot be perfectly converted into useful work. Friction, vibration, and heat loss result in energy leakage. But even a frictionless heat engine perfectly insulated against heat loss would still be unable to transform all its energy input into work.

The First and Second Laws of Thermodynamics[78] explain why this is so. There are some complex mathematics behind each of these laws, but they can be roughly summarized as follows:

The First Law

Energy is conserved—it can neither be created nor destroyed. That is, energy input must equal the total energy output; the input must equal the sum of useful work produced, friction loss, heat loss, etc.

Energy can be transformed from one form to another (i.e., potential to kinetic or kinetic to potential) any number of times. During each such transformation, some energy may be lost into the environment, but the converted energy plus the energy lost must equal the original amount of energy in the system. The total energy in the system remains the same.

The Second Law

Energy flows "downhill." Objects fall down, not up; heat flows from hot objects to cold; and fluids and gases flow from high pressure to low.

---

[78]As its name implies, *thermodynamics* is the study of heat in motion.

Machines need *energy differentials* in order to work. Engines must operate between high and low temperature *reservoirs*, high and low pressure zones, or high and low *electric voltages*.

When energy flows downhill, the system's total energy differential is reduced. Energy becomes more evenly distributed and less available to perform useful work. Imagine, for example, a waterfall that pours into the ocean. Suppose that the falling water is used to turn a waterwheel and produce work. When the water reaches the ocean, it still has potential energy because it is above the center of the Earth. Yet, this energy cannot be exploited because there is no lower elevation to which the water can flow.

We can use machines to reverse the direction in which energy naturally flows. Pumps push water uphill, refrigerators force heat to flow from cold to hot, and compressors drive gases from low to high pressure. However, the system's *overall* energy differential must be reduced even when a machine is used to reverse the natural flow.

A system's *entropy* can be thought of as a measure of the evenness of its energy distribution. The higher a system's entropy, the less available is its energy to do work or to drive machinery. During any energy conversion, the entropy of the entire system must increase. While expending energy can bring order to a subsystem, the disorder of the total system must increase.

Given the Second Law, it is often asked how the Earth has become more ordered (with, for example, the appearance of ever more complex life forms). The answer is that the Earth is not a closed system. While the entropy of the universe (the total system) is always increasing, there is a constant and tremendous flow of energy from the sun to the Earth (a subsystem).

According to Isaac Asimov, "The Earth receives only one-half of one-billionth of the sun's radiant energy. But in just a few days it gets as much heat and light as could be produced by burning all the oil, coal, and wood on the planet."[79]

Because energy can be transformed but not destroyed (the First Law) the Universe will never run out of energy. However, the Second Law dictates that eventually the energy will no longer be usable. That point would come when the Universe is at a completely

---

[79]Isaac Asimov, *Isaac Asimov's Book of Facts* (New York: Wings Books, 1979), p. 108.

uniform state and there are no longer any energy differentials. With no differentials, no useful work can be done and all life must end.

Ultimately, Thomas Malthus and his gloomy predictions may be correct—we are all doomed. However, the Universe's death by entropy, if it occurs, is many billions of years in the future, so his last laugh will be a long time in coming.

On the other hand, entropy may ultimately be conquered by gravity—the attractive force between matter. If the mass of the Universe is sufficient, gravity will cause it to collapse back to a single point, and another Big Bang could start the whole thing all over again. The Universe may alternately expand and collapse indefinitely and may have already done so countless times.

EFFICIENCY

Most of the energy that goes into producing electricity is lost. Say that a utility company's power plant runs on coal. The coal is burned to boil water and produce steam that, in turn, drives a turbine. The turbine runs a generator, and the generator produces electricity. At each step, energy is lost.

When the coal is burned, some of the heat that is produced escapes up the smokestack along with the hot gases that are formed. At the next stage, less than half of the steam's energy is actually used in driving the turbine. Most is lost to the atmosphere when the spent steam leaves the turbine.

Next, because of heat and friction losses, the generator is unable to convert all of the turbine's kinetic energy into electricity. Then there are power losses in the transmission lines that carry the electricity.

Finally, when the electricity is used, there are still more energy losses because appliances are unable to convert all of their power input into useful work. Throughout this entire process, most of the energy originally stored in the coal is lost. Only a small fraction of it actually goes for productive work.

The *efficiency* of a given machine is defined as the ratio between the usable work that comes out of the machine to the energy that went in. Many things can be done to improve efficiency. For example, insulation can be added to slow heat loss, and lubricants can reduce friction. But, again, no machine can be made to be perfectly efficient. The following chart shows some typical efficiencies of various devices.

From the next table, it can be seen that only about 25 percent of the energy in a gallon of gasoline is actually used to move a car, while the other 75 percent is lost. Or, take the example of the power plant mentioned above. If the boiler has an efficiency of 85 percent, the steam turbine 45 percent, and the generator 95 percent, then the efficiency of the overall system is $0.85 \times 0.45 \times 0.95$ or about 36 percent. If that electricity is used to turn on a light bulb, then the overall efficiency of converting coal into light is approximately 0.36×0.05, or less than 2 percent!

While an electric heater is able to turn all of its power input into heat, a gas furnace is still more efficient if the entire system (from power generation

Energy Conversion Device	Energy Conversion	Efficiency (%)
Electric heater	Electricity/Thermal	100
Electric generator	Mechanical/Electrical	95
Electric motor (large)	Electricity/Mechanical	90
Battery (dry cell)	Chemical/Electrical	90
Steam boiler (power plant)	Chemical/Thermal	85
Home gas furnace	Chemical/Thermal	85
Home oil furnace	Chemical/Thermal	65
Electric motor (small)	Electrical/Mechanical	65
Natural gas combined cycle	Chemical/Mechanical	60
Home coal furnace	Chemical/Thermal	55
Steam turbine	Thermal/Mechanical	45
Diesel engine[80]	Chemical/Mechanical	43
Gas turbine (aircraft)	Chemical/Mechanical	35
Gas turbine (industrial)	Chemical/Mechanical	30
Automobile engine	Chemical/Mechanical	25
Fluorescent lamp	Electrical/Light	20
Human[81]	Chemical/Mechanical	18
Silicon solar cell	Solar/Electrical	15
Steam locomotive	Chemical/Mechanical	10
Horse[82]	Chemical/Mechanical	10
Incandescent light (light bulb)	Electrical/Light	5

Source: Pennsylvania State University's Earth and Mineral Sciences web site: www.ems.psu.edu/
~radovic/fundamentals4.html, unless otherwise noted.

to consumption) is considered. Remember that the power plant's efficiency is only about 36 percent. Because the electric heater's efficiency is nearly 100 percent, the efficiency of the overall "power plant/heater" system is 0.36 × 1.00 or 36 percent. That is still much less than a gas furnace's efficiency of 85 percent.

In other words, it is far more efficient to burn natural gas in a house to produce heat than to burn it at a power plant in order to produce electricity that will later be converted into heat at the house.

ENERGY ECONOMICS

Throughout this chapter, the relative costs of the various methods of converting energy into useful work have been emphasized. Some believe that the most environmentally benign technologies should always be adopted regardless of

---

[80]Robert Brady, "Diesel Cycle Engines" in John Zumerchik, ed., *Macmillan Encyclopedia of Energy* (New York: Macmillan Reference, 2001), p. 333.

[81]J. R. McNeill, *Something New Under the Sun*, p. 11.

[82]Ibid.

cost. This view fails to recognize that the cost of producing something reflects, to some degree, the effort, resources, and pollution that were spent to make it.

> **❝**Economics is the study of how individuals transform natural resources into final products and services that people can use.**❞**[83]
>
> **Mark Skousen**

Market prices make it possible to keep score—that is, they enable us to compare the relative value of different resources and decide whether a given action is worthwhile. For example, if the efforts of an oil producer are to be of any use, they must produce more energy in the form of oil than is expended in order to recover and refine that oil. The activity must, in effect, make an energy profit. But how can a producer know whether a net profit is being made?

Suppose an oil company discovers a well that it estimates will produce 100 barrels of oil a day. Should the company produce the oil or cap the well? To decide, it could perform an energy balance comparing the number of BTUs contained in the oil against the energy needed to produce and process it. Unfortunately, doing this calculation would be nearly impossible. The company would have to determine the energy required to mine iron ore, transform it into steel, shape the steel into pumps, pipes, valves, and bolts. It would also have to know the energy used to transport this equipment to the well site and install it. Similarly, the energy needed to create, transport, and install all the other materials used would have to be calculated along with that consumed by the laborers and their families while the work is under way.

Even if the company somehow determines that producing the well would result in a net energy gain, what then? Should all 100 barrels a day be produced? The energy balance does not indicate consumer demand. If we only need 50 barrels worth of energy a day, will pumping all 100 barrels leave us twice as well off or just leave us with a storage problem?

An energy balance has another shortcoming. The only reason we could even consider determining profit or loss by comparing energy expended against energy produced is that energy appears on both sides of the equation. It would be reasonable, for example, to invest 50 BTUs in order to recover 100 BTUs worth of oil. But how much energy should be expended to produce a pound of copper?

In a free market, *prices tend, over time, to reflect the costs of producing a commodity.* The oil producer does not need to know how much energy it takes to build a pump; he needs to know only the pump's price. Included in this price are the pump manufacturer's costs for overhead, labor, materials, and energy.

---

[83]Mark Skousen, *Economics on Trial*, p. 18.

Corbis

Knowing the price of the equipment, the cost of its transportation and installation, and the price that consumers are willing to pay for his products, the producer can calculate the monetary profit that he would receive by recovering the oil.

As long as they make a monetary profit, then, producers can be reasonably sure that they are also making a net energy profit.[84]

MARKET PRICING

In addition to production costs, market prices also reflect demand. If a cold spell in one area of the country increases demand for heating oil, its price will rise, and producers will send more oil there to maximize their profits. As prices drop due to increased supply, oil will be shifted to other markets.

Prices also allow the relative values of different goods to be compared at any moment. This is critical information since at any given time resources are limited. Relative prices tell manufacturers what people value most and therefore what they should use their resources to make. Producers that supply the public with the goods they want at prices they are willing to pay will

[84]Viewed in this light, profits perform an essential social service. They provide a signal that indicates whether resources are being efficiently used to provide for consumer needs and desires. Companies that employ the fewest resources to best satisfy customers will make profits. Their success will attract both investors and imitators. Those that provide the least amount of satisfaction at the highest cost will lose money, and either go out of business or change their ways.

make profits. Thus, the market automatically directs more resources to those producers that best meet consumers' needs.

> The phrase, "the market," does not refer to some vast, impersonal, institution that controls individuals and corporations. It refers instead to the continuous exchange of goods, services, and ideas by millions of individuals—some acting on their own behalf and others on the behalf of companies and institutions. These countless actions make up the market. In a *free* market, people communicate, buy, sell, trade, and otherwise interact without third-party coercion (i.e., use or threat of force).

Oil refineries provide a good example of how price drives production. A refinery can turn a barrel of oil into a number of products, including gasoline, diesel, heating oil, lubricants, and feedstock for plastics. By adjusting the refining processes, production can be shifted to make more of one product and less of another. Refiners continually monitor the market prices of the products they make so that they can adjust their output in response to shifts in consumer demand. By so doing, they satisfy their customers and maximize their profits.

A government-run refining monopoly, by contrast, would be driven by politics rather than by consumer demand (as indicated by market price). If, for instance, the farm lobby is particularly powerful, the directors of a government-run facility would hesitate to offend that lobby by shifting production away from the diesel needed by farm equipment and towards another product like gasoline.

THE BIG PICTURE

One of the powers of the free market pricing system is that it incorporates the big picture into local decision making. For example, recycling is typically presented as inherently good—something so obviously beneficial as to be beyond question. But recycling not only saves resources, it also *costs* resources. Recycling plants must be constructed. Used materials must be separated, collected, transported to the plants, and processed. If recycling a ton of paper costs more resources and produces more pollution than it saves, why do it? Without a free-market pricing system, the environmental impact of recycling cannot be determined.

If local energy efficiencies were the only thing that mattered, we would tear down and replace the country's power plants every time more efficient technology became available. While this would ensure that our power plants would always convert energy resources into electricity as efficiently as possible, overall, resources would be wasted.

> Economist F. A. Hayek described the price system as "a mechanism for communicating information" whereby dispersed and fragmented bits of information are brought together into a rational whole. "The whole acts as one market, not because any of its members survey the whole field, but because their limited individual fields of vision sufficiently overlap [through relative prices] so that through many intermediaries the relevant information is communicated to all."[85]

A classic example is the creation of the American transcontinental railway in the 1860s. While it was being built, there was continuous debate between the engineers and the financiers. The engineers wanted to use the best construction techniques and the most durable building materials available. The financiers, on the other hand, were deeply in debt and wanted the railroad built as quickly as possible so that it could start generating income. In the end, the financiers won. As a result, much of the railroad's infrastructure had to be rebuilt within a few years of its construction. Railroad ties made from green timber and bridges built of wood rather than stone all had to be replaced.

A tragic waste? Perhaps not. Consider how many resources were saved by the railroad's existence. Goods no longer had to travel by ox cart or by ship around South America. Travel times were cut from months to mere days. Moreover, once the railroad was in place, the construction materials and workers needed to rebuild the railroad could be transported much more quickly and efficiently than before. In all likelihood, completing the railroad sooner rather than later saved far more resources than it wasted.[86]

It may well be, then, that the trade-off between *economic efficiency* and *resource efficiency*, is not a trade-off at all. If a sufficiently encompassing resource balance is made, conserving money should translate into conserving resources. This should not be surprising as money is used to purchase resources either directly (goods) or indirectly (services). While price distortions could sever the relationship between money and resources, such distortions are typically the result of government interference with the marketplace (e.g., currency inflation, or price controls).

> William Stanley Jevons founded the study of energy economics with his 1865 book, *The Coal Question: An Inquiry Concerning the Progress of the Nation, and the Probable Exhaustion of Our Coal Mines*. Jevons warned that England's coal boom was coming to an end, and her industry would migrate abroad (to America in particular) where energy supplies were more plentiful. The ensuing "coal panic" caused Parliament to consider retiring the national debt to help the country weather the expected energy crisis.

---

[85]F. A. Hayek, "The Use of Knowledge in Society," *Individualism and Economic Order* (Chicago: Henry Regnery, 1948), p. 86.

[86]For more on the building of the transcontinental railroad, refer to Stephen Ambrose, *Nothing Like it in the World* (New York: Simon and Schuster, 2000).

But the coal famine did not come. Improvements in mining technology kept costs steady. Oil and, later, natural gas came into the picture, two energy substitutes that Jevons scarcely considered.

To his credit, he understood the *economic challenge* as "the gradual deepening of our coal mines and the increased price of fuel," not that "our coal seams will be found emptied to the bottom, and swept clean like a coal-cellar."[87] Supply would not run out, it would just become more expensive.

Could renewable energies fuel England's industrialization? Jevons was pessimistic. "The wind," he argued, "is wholly inapplicable to a system of machine labour, for during a calm season the whole business of the country would be thrown out of gear."[88] Regarding waterpower, "In very few places do we find water power free from occasional failure by drought."[89] What about burning wood? "We cannot revert to timber fuel," he stated, for "nearly the entire surface of our island would be required to grow timber sufficient for the consumption of the iron manufacture alone."[90] Geothermal? "The internal heat of the earth . . . presents an immense store of force, but, being manifested only in the hot-spring, the volcano, or the warm mine, it is evidently not available."[91]

Nearly a century-and-a-half later, Jevons' concerns remain relevant to the energy sustainability debate. Intermittency, variability, and (un)availability are still obstacles to a significantly increased role for renewable energies in today's economy.

INSTITUTIONS AND ENERGY

Industry requires land, labor, and capital.[92] But these are not enough. Indeed, most third world countries possess all these physical ingredients of thriving industries. What they lack are the institutions that give life to markets. Property rights and the rule of law provide the framework that allows people to turn dead material into the stuff of life. We in the West are so accustomed to these institutions that we no longer see them—they are like the air we breathe. Yet without them our civilization and perhaps even our technology would be impossible.

Property rights are a precondition for trade; one cannot sell that which one does not own. The concept of ownership, then, is fundamental to markets. Nearly as important is the ability to identify a thing's owner. A would-be purchaser of an object must be able to establish that he or she is, in fact, dealing

---

[87]William Stanley Jevons, *The Coal Question: An Inquiry Concerning the Progress of the Nation, and the Probable Exhaustion of Our Coal Mines*, introduction to the second edition, Macmillan, 1866, reprinted in A. W. Flux, ed., *The Coal Question* (London: Macmillan, 1906), p. xxix.

[88]Ibid., first edition, 1865, p. 122.

[89]Ibid., p. 129.

[90]Ibid., p. 140.

[91]Ibid., p. 120.

[92]Economists define "capital" as physical assets such as natural resources, factories, ships, trucks, or roads.

with the object's owner. Such proof is provided by deeds and bills of sale recognized under the rule of law. Such pieces of paper are abstract or symbolic representations of physical things. Often when a thing, such as a plot of land or a building, is bought and sold, it is only this paper representation that actually changes hands. People in a society recognize that ownership has been transferred through such representational means in accordance with the laws of the land.

Abstract representations of physical objects do far more than just enable them to be bought and sold. They provide a sort of institutional trust that allows buyers and sellers to trade with the confidence that the terms of their agreement will be fulfilled, and, if necessary, enforced. Your home address, for example, is nothing more than a symbolic representation of an actual place, yet it provides a means of locating and identifying your residence. It establishes a point of contact and allows you to receive merchandise, services, and invoices for them at your home. Utility companies can find your house in order to deliver electricity, phone service, water, and gas, and be assured of receiving payment in return.

Documented property can also be used as collateral to borrow money to start a new business or to pay for college tuition. Without legally recognized deeds, capital is idle and its potential wasted.

Such institutions don't just happen, they evolve, as Hernando de Soto explains in his book, *The Mystery of Capital*.[93] Even though one of the founding principles of the United States was respect for each individual's property, many years passed before such rights were codified.

> **❝***The first and chief design of every system of government is to maintain justice; to prevent the members of a society from incroaching on one anothers property, or seizing what is not their own.***❞**[94]
>
> **Adam Smith**

In America's early years, it was not uncommon for a piece of land to be claimed by any number of people—a squatter who built a crude cabin on it and began farming; a trapper who had purchased hunting rights from local natives; a soldier who had been awarded the land by the government he had served; or a railroad that had been given land grants to encourage it to lay

---

[93]Hernando de Soto, *The Mystery of Capital: Why Capitalism Triumphs in the West and Fails Everywhere Else* (New York: Basic Books, 2000), pp. 105–151.

[94]Adam Smith, *Lectures on Jurisprudence* (Indianapolis, Liberty Press, 1762, 1978), p. 5.

track. Disputes over such lands were fierce and sometimes bloody. Occasionally, troops were called in to run squatters off. They burned cabins, broke fences down, and destroyed crops. But often the people came back and rebuilt their homes as soon as the soldiers had gone.

In the absence of institutional resolutions to conflicting claims, squatters—the people on the ground—developed their own *extra-legal* systems. They wrote up deeds that were recognized by others in the community. Eventually, many such extra-legal property rights were absorbed into law, but the nineteenth century was nearly over before the rule of law caught up with a growing and evolving nation.

The United States was fortunate that property rights were fairly well established (at least in principle if not always in fact) by the time Edwin Drake drilled his oil well in 1859.

> Petroleum production presented new property rights issues because oil (and natural gas), unlike other minerals, can migrate from one part of a reservoir to another. Eventually a system evolved in which surface and subsurface rights were held to be distinct (that is, the surface rights to a piece of land can be bought and sold independently of the subsurface, or mineral rights).
>
> In addition, U.S. Courts established the *rule of capture* that held that oil was owned by whoever pumped it out of the ground.[95] This rule created an incentive for different producers owning or leasing land lying atop the same reservoir to sink wells and pump out the oil as fast as possible. Such rapid production wasted resources and resulted in reservoir damage, reducing the amount of oil that could ultimately be recovered from the field.[96]
>
> The problems caused by the rule of capture were eventually solved when companies *unitized* their fields, i.e., let one company handle production while all shared the costs and profits according to a negotiated formula. Regulatory roadblocks and fear of anti-trust suits, however, delayed this solution for years.[97]

Despite problems with American property rights laws as they applied to oil and gas reservoirs, the laws provided the necessary framework by which these resources could be found, produced, and sold.

---

[95] In the 1880s, the Pennsylvania Supreme Court established the "rule of capture" granting property rights to "migrant" minerals such as petroleum to the act of physical possession (as versus the ownership-in-place law for hard minerals).

[96] For more on the problems associated with U.S property rights as established for petroleum reserves, see Robert Bradley, Jr., *Oil, Gas & Government: The U.S. Experience* (Lanham: Rowman & Littlefield, 1996), pp. 64–69.

[97] Ibid., pp. 109–31.

Within such a framework, oil production benefited the entire population of the country. Farmers and ranchers, under whose land the oil was found, were compensated for the use of their land by oil explorers, drillers, and producers. The owners, employees, and shareholders of oil companies, and of oil well service and supply companies, benefited. Refiners and retailers profited as well. Most of the benefits, however, accrued to the millions of people who were supplied with increasingly affordable energy to power their cars, homes, places of business, towns, and cities.

People living in countries without strong private property laws benefit far less from the land's oil and mineral wealth. In the first place, companies hesitate to invest in countries that do not recognize property rights. Oil production and mining are *hardware* businesses that require a lot of machinery, concrete, and steel. How can you build a facility if you cannot establish clear title to land on which to build it? Why build a plant or factory if the local government can lay claim and take it from you? As Lee Raymond, the chairman of ExxonMobil, the world's largest oil company, stated, "We are prepared to take the commercial risks that accompany the fluctuations of the world energy market. But we do not want to take unnecessary legal risks, especially those that arise from deficiencies in the legal structure of a country."[98]

Outside North America and Europe, government ownership of subsurface rights is the rule rather than the exception. In fact, governments control most of the world's oil and gas reserves—from the Middle East, to countries of the former Soviet Union, to Central and South America. Often governments offer *concession agreements* to outside companies in exchange for developing their resources. Such agreements establish unambiguous legal title and clearly spell out responsibilities. The parties (i.e., the government and the production companies) agree upon a framework for legally binding dispute resolution and guarantee restitution in case of damages. These agreements offer companies an element of security to offset the risk of investing in a country in which laws follow the whims of the rulers.

Even under such arrangements the citizens of developing countries do not always benefit from the production of oil and minerals to the same degree as do citizens of western nations. All too often, "public ownership" translates into ownership by the rulers. Money paid to a government by oil producing companies may go toward building the nation's infrastructure, or it may just disappear into the rulers' personal offshore bank accounts.

[98]Lee Raymond, "The Rule of Law in International Oil & Gas Development," *World Energy*, Vol. 5, No. 1 (2002), p. 111.

Corbis

Governments in poor countries do both too much and too little. South American economist Hernando de Soto and his team of researchers worked for 289 days to get all the certifications needed to open a small garment workshop on the outskirts of Lima, Peru. "The cost of legal registration was $1,231—thirty-nine times the monthly minimum wage."[99] Some 26 months of effort were needed for a taxi driver to get approval for a route. Obtaining permission to build a house on state-owned land took nearly seven years and 207 administrative steps by 52 government offices. De Soto's team found similar levels of red tape in Egypt and Haiti.[100]

At the same time that these governments have erected monumental roadblocks to individual initiative and productivity, they have failed to create the legal structure essential to any modern society. What developing nations need most are legal systems that document and uphold property rights, enabling people to easily transfer, trade, and borrow against their property.[101]

THE ENERGY INDUSTRY

The petroleum industry is actually made up of five sub-industries or *sectors*. Each sector represents a different stage in the petroleum processing chain. In industry jargon, the exploration and production sector is called the *upstream* part of the business, transportation the *mid-stream*, and refining, wholesaling, and retailing the *downstream*.

---

[99]Hernando de Soto, *The Mystery of Capital*, pp. 19–20.

[100]Ibid., pp. 20–21.

[101]Ibid., pp. 210–11. Also see the box discussion on page 189 below.

The largest firms in the energy industry are *integrated* across these industry sectors. *Oil majors* such as ExxonMobil and Royal Dutch Shell produce, refine, transport, and market crude oil and oil products.

Oil becomes increasingly valuable as it moves downstream. Yet the supply of, and consumer demand for, the final product determine the value of the entire chain of activities, not the other way around. Crude oil's value is governed by the prices consumers pay for gasoline, fuel oil, and other end products, rather than what crude oil costs to find, produce, and refine.

Over time, however, costs and prices tend to approach each other. In order to stay in business, a company must charge enough for its product to cover its costs. On the other hand, if a firm charges substantially more than its costs, competitors will move in and win customers by offering lower prices.

OIL INDUSTRY SEGMENTATION

The petroleum industry is divided into five general sectors from the production of crude oil to the final sale of petroleum products. Each of these divisions has associated service industries such as drilling contractors for exploration and production and pipeline construction companies for transporters.

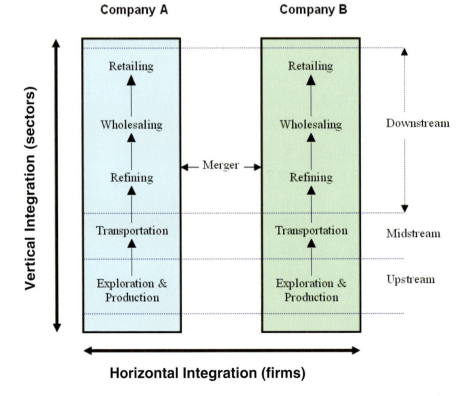

Shown in the illustration are two forms of corporate integration. *Horizontal integration* is expansion by a company within its own sector, perhaps by purchasing or merging with a rival. For example, the "seven sisters" of the 1960s oil industry, Exxon, Gulf, Texaco, Shell, Chevron, Mobil, and British Petroleum, are now four: ExxonMobil, Shell, BP, and ChevronTexaco (Chevron combined with Gulf before its merger with Texaco).

The entrance of a company into a new sector of the production chain—for example, a producer integrating forward into refining or a refiner integrating backward into production—is termed *vertical integration*. Vertically integrated companies can control risk and quality throughout the processing chain, a strategy that made John D. Rockefeller's Standard Oil so successful.

Many smaller *independents* are able to compete with the large integrated firms, especially in niche markets. Shifting consumer demand determines the market shares held by integrated, partially integrated, and non-integrated (independent) energy firms. The presence or absence of *economies of scale* (falling costs from larger output) and *economies of scope* (falling costs from performing more than one function) determine the size and structure of firms in a free market.

Globally, petroleum companies include both privately-owned *capitalistic* firms and government-owned *socialistic* firms. Capitalism is the dominant economic system in the United States energy market with some exceptions.[102] In Mexico, on the other hand, one giant government-owned company, Petroleos Mexicanos (PEMEX), has a legal monopoly over all oil- and gas-related functions. Other state-owned monopolies include Petroleos de Venezuela, S.A. and

Corbis

---

[102]U.S. municipalities own over a hundred entities that distribute natural gas or electricity in their jurisdictions. The U.S. Department of Energy, a federal agency, operates hydroelectric facilities that were built by the government decades ago.

Repsol in Spain. In such countries, as explained in the last section, the government not only owns the means of production but the oil reservoirs themselves.

In the United States, the natural gas and electricity industries each have three segments—production, transmission, and distribution. Regulation has played a large role in determining the structure of both of these industries. In the past, electric utilities were completely vertically integrated, from production, to transmission, to marketing, while the natural gas industry was totally non-integrated. Relaxed regulation is likely to make the structure of these energy industries more like that of petroleum—a blend of integrated and non-integrated firms. This trend has already begun as some electric utilities have sold their generation facilities to independent power producers.

To the untrained eye, the energy industry may seem like a collection of physical resources: petroleum reservoirs and mineral deposits, oil wells and mines, refineries and power plants, pipelines and power lines. But *intellectual capital* drives these physical assets. The *entrepreneurial* component of the energy business, in which new technologies and strategies are employed to do entirely new things or perform old tasks in new ways, is the engine of progress described in this book. Economist Joseph Schumpeter described capitalist progress as *creative destruction*, a process in which new techniques and technologies render existing modes of operation obsolete.[103]

ECONOMICS AND POWER CONSUMPTION

The amount of electricity used in any community varies throughout the day. Consumption is typically much greater during daylight hours than at night when most people are asleep. Usage also varies by season. Much more is needed in summer to run air conditioners and in winter to power heaters than in either spring or fall. But regardless of the time of day or the season, when consumers turn on a switch, they expect the power to be there.

In order to meet this uneven demand, power companies must build their plants large enough to handle peak loads. As more people and industry move into an area, however, the companies may find that their facilities can no longer handle the new demands being placed on them. When this happens, should the old plants be expanded, or should new plants be built?

An alternative to either of these options is to use the existing plant's excess capacity during off-hours to generate power into storage. Then, during times of high demand, this storage can be tapped to supplement the main generators.

---

[103]Joseph Schumpeter, *Capitalism, Socialism and Democracy* (New York: Harper & Row, 1942, 1962), p. 83.

One way to store excess energy is to pump water up into a reservoir. When additional electricity is needed, the pumps are turned off, and water is released through the dam to turn hydroelectric turbines. A future method might be to use off-peak electricity to break water molecules into oxygen and hydrogen atoms. Hydrogen is an exceptionally efficient and clean burning fuel and can be used to produce electricity.

The problem with such schemes is that, while they may reduce the resources needed to build a new or bigger power plant,[104] they result in higher overall fuel costs. Fuel must be burned to produce the electricity needed to pump the water up into the reservoir. As explained in the discussion of efficiency, significant amounts of energy will be wasted in the process of converting electrical power into potential energy (the increase in the water's energy by virtue of its being placed at a higher elevation). Then only some of the potential energy will be regained when the water falls back down through the turbines. Because of the costs involved, such storage techniques are the exception.

Perhaps the best way to reduce the need for new capacity is to increase the price of power used during peak hours. These higher prices would encourage customers to shift consumption from critical times to periods when demand is lower. People might, for example, choose to run their clothes washers and dryers at night when rates are lower. New metering technology is making this possible.

Nothing is free, and there are no perfect solutions. There are always trade-offs; one thing is lost in order to gain another. Market incentives lead people to balance these trade-offs to make the best use of available resources.

---

[104]Assuming, of course, that the cost of building the facilities necessary to store and reuse the old plant's excess power is less than that of expanding the old plant or building a new one.

WILL WE RUN OUT OF ENERGY?

CHAPTER

4

T HE PERILS OF PREDICTION

In 1865, economist William Stanley Jevons wrote a book warning that the coal supply that had helped make the United Kingdom an economic power was rapidly depleting.[105]

David White, Chief Geologist for the U.S. Geological Survey, said in 1919, "the peak of [American oil] production will soon be passed—possibly within three years."[106]

In a 1968 bestseller, author Paul Ehrlich proclaimed, "The battle to feed humanity is over. In the 1970s and 1980s hundreds of millions of people will starve to death in spite of any crash programs embarked upon now."[107]

The Limits to Growth, published in 1972 by the Club of Rome, calculated that given then current trends, the world could be out of petroleum by 1992 and natural gas by 1993.[108]

Two years later, Ehrlich and his wife, Anne, stated, "we can be reasonably sure . . . that within the next quarter of a century mankind will be looking elsewhere than in oil wells for its main source of energy."[109] Their Malthusianism was also captured in the equation, I = PAT, where negative environmental *impact* was directly proportional to increasing *population, affluence,* and *technology.*[110]

---

[105]William Stanley Jevons, *The Coal Question.*

[106]Quoted in Edward Porter, *Reinventing Energy: Making the Right Choices* (Washington: American Petroleum Institute, 1995), p. 17.

[107]Paul Ehrlich, *The Population Bomb* (New York: Ballantine Books, 1968), p. xi.

[108]Dennis Meadows and Donella Meadows, *The Limits to Growth* (New York: Universe Books, 1972), p. 193.

[109]Paul Ehrlich and Anne Ehrlich, *The End of Affluence* (Rivercity, MA: Rivercity Press, 1974), p. 49.

[110]Paul Ehrlich, Anne Ehrlich, and John Holdren, *Human Ecology: Problems and Solutions* (San Francisco: W. H. Freeman Company, 1973), chapter 7.

In 1979, James Schlesinger, the first Secretary of the U.S. Department of Energy and a Ph.D. economist, stated, "The energy future is bleak and is likely to grow bleaker in the decade ahead."[111] He was not staking out a radical position, but instead was merely repeating the conventional wisdom of the day.

Not only were policy makers convinced that oil was running out, but many top executives in the petroleum industry accepted it as well. In fact, in the late 1970s and early 1980s some major oil companies began shifting resources out of the oil business. Exxon purchased Reliant Electric, a manufacturing firm, and started up an office equipment company; Gulf Oil Company opened a uranium mine in northwestern New Mexico; ARCO bought Anaconda, a large mining concern; SOHIO purchased Kennecott, another mining company; and Mobil bought the Montgomery Ward retail chain.[112]

Since hydrocarbon production began, many smart people have made thousands of pessimistic predictions—each just as alarming and each just as wrong. Not only has the world's known supply of coal and hydrocarbons failed to disappear, it has actually grown—substantially! The proof of this lies in the fact that, despite temporary price spikes such as those experienced in 2004, finding costs and selling prices have declined over the long run. If fuel were becoming scarcer, its cost would be increasing.[113]

In addition to comparing inflation-adjusted prices over time, changes in scarcity can be measured by comparing the amount of labor time that an average worker needs to expend to earn the income to purchase a particular item. Major forms of energy have grown substantially cheaper measured in *work-time pricing* as seen in the figure on the following page.

So what happened? How could all of these people's predictions have been so mistaken? First, let's look at how they came up with their numbers. Most simply took the amount of *known recoverable reserves* of a particular resource and divided it by the amount that was being used each year. The result of the division was the number of years left, or *reserve years*. For example, suppose that a nation's geologists have located 200 million tons of coal, and that the country is using coal at a rate of ten million tons per year. They should expect to run out of coal in twenty years. How could there be problems with such a simple calculation?

The first problem is with the word "known" in the phrase "known recoverable reserves." What is known is constantly changing. People are continually searching for and finding more resources.

Exploration is expensive. It takes a lot of people, equipment, and time to find oil, coal, natural gas, and so on. Because exploration is so costly, it does

---

[111]Quoted in Mark Mills, *Getting It Wrong: Energy Forecasts and the End-of-Technology Mindset*, Competitive Enterprise Institute, 1999, p. 12.

[112]Joel Darmstadter, Hans Landsberg, Herbert Morton, and Michael Coda, *Energy Today and Tomorrow: Living with Uncertainty* (Englewood Cliffs, NJ: Prentice-Hall, 1983), p. 13.

[113]Also see the graph on p. 50 on inflation-adjusted gasoline prices.

DECLINING WORK-TIME PRICE OF U.S. ENERGIES

Energy cost in terms of work time is the best measure of affordability. Today, the average laborer can buy a week's worth of gasoline and electricity for about 90 minutes of work. The same amount of energy cost a full workday in the 1920s. *Source: Myths of Rich & Poor* by Michael Cox. Copyright 1999 by Michael Cox and Richard Alm. Reprinted by permission of Basic Books, a member of Perseus Books, L.L.C.

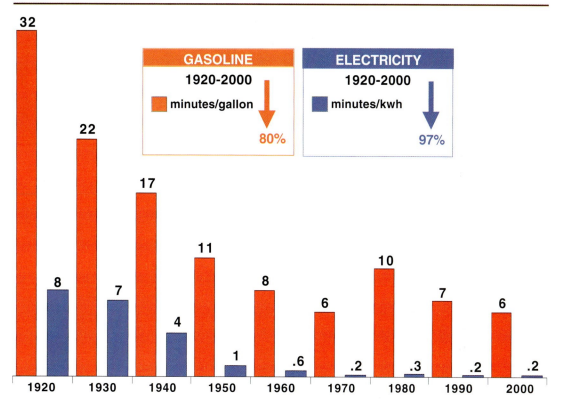

not make much sense to look for resources that will not be needed for two or three hundred years. That is why the known reserves of so many resources often seem to fall in the range of fifteen to twenty years regardless of how many years have passed or how much of the resources have already been produced.

The next problem is with the word "recoverable." What is really meant by this word is *economically* recoverable. No one is going to spend, say, $100 to dig up a ton of coal if he can get only $50 for it. Suppose, though, that the price goes up to $150 per ton. Suddenly, that coal becomes worth mining. When prices change, the amount of reserves that can be recovered economically also changes. In addition, as the timeline (*Appendix* A) indicates, people often come up with better and more efficient ways of doing things. Suppose that someone invents a less expensive way to mine coal so that it now costs only $25 per ton to dig. Even without a price increase, the coal is now worth producing.

U.S. Coast Guard

Finally, there is another problem with the simple "number-of-years-left" formula—the annual consumption rate. When conditions change, people's actions change. If in the future natural gas became harder to find and produce, its price would rise, and the higher cost would encourage people to use energy more efficiently or simply use less altogether. Instead of keeping their homes so warm in the winter, maybe they would start wearing a sweater around the house. Or perhaps they would find substitutes for gas, such as petroleum from oil shale or tar sands. As these resources wane, fuel can be made from plant matter such as algae, seeds, and vegetable oils.

It is a mistake to confuse a *resource* with the *service* it provides. People want their homes to be warm and comfortable, but they do not really care whether it is done by burning coal or by splitting atoms. They want to be able to make a phone call; whether the call goes over thousands of tons of copper wire, is sent over a strand of fiber optics, or is beamed through the air by microwave is of no concern.

> **❝**Perhaps the very concept of exhaustible reserves ought to be discarded as wrong or irrelevant. Not much of the resources we know today will ever be used because better ones will be found. Or the need itself may disappear before the resource.**❞**[114]

M. A. Adelman

---

[114]M. A. Adelman, "My Education in Mineral (Especially Oil) Economics," *Annual Review of Energy and the Environment*, vol. 22 (1997), p. 26.

THE ULTIMATE RESOURCE

What is left out of the "number-of-years-left" equation is *human ingenuity*. People discover new and better methods of finding resources. They learn ways to conserve, and they find substitutes. Often people even find uses for substances that, before, no one ever thought of as resources. If energy is the *master resource*, then creative and knowledgeable people are the *ultimate resource*.[115] Economists like Julian Simon argue that as long as people are free to use their minds and to act upon their ideas, the world will never run out of energy. Resources spring from knowledge, not the ground.[116]

A common mistake that many make is to project current trends into the future as if they will continue forever. The absurdity of making such assumptions can be demonstrated by a simple example. Suppose that a ten-pound baby doubles her weight in her first year of life. What reasonable person would become concerned by the fact that, if the child's growth continued at the same rate (i.e., doubling every year), by age 10 she would weigh more than 5,000 pounds? Obviously, children do not keep growing at the same rate and the rate of change declines without any intervening catastrophe.

This is a far-fetched example, but consider the following actual occurrence. At the turn of the century, engineers warned that if demand for electrical power in the Chicago metropolitan area continued growing at then current rates, the city's entire inner loop would be covered by power plants within a few years. In fact, the rate of growth *increased*, but Chicago was "saved" by a technological breakthrough. Reciprocating steam engines (which had been powering the city's generators) were replaced by turbines with much higher power-to-size ratios.[117]

When Malthus looked at people, he saw mouths that must be fed and forgot the minds behind the mouths.[118]

> **"**The evidence suggests that over the past century, new technology driven by free market forces has overcome the geophysical scarcity of nonrenewable natural resources. Increased reliance on markets during the closing decades of the twentieth century is cause for optimism that these trends will continue in the twenty-first.**"**[119]

Stephen Brown and Daniel Wolk

---

[115]Julian Simon named his book, *The Ultimate Resource*, to make exactly this point.

[116]Erich Zimmermann, *World Resources and Industries* (New York: Harper & Brothers, 1951), p. 10.

[117]Forrest McDonald, *Insull*, p. 98.

[118]Paraphrased from Peter Huber, *Hard Green* (New York: Basic Books, 1999), p. 10.

[119]Stephen Brown and Daniel Wolk, "Natural Resource Scarcity and Technological Change," *Federal Reserve Bank of Dallas—Economic and Financial Review*, 1st Quarter, 2000, p. 9.

DISSENTING VOICES

Pessimists like Paul and Anne Ehrlich admit that advances in exploration technology and fuel efficiency have stretched world fossil fuel supplies far more than they had predicted. However, they argue that no matter how long supplies last, there must eventually be a point at which they can no longer be economically extracted. Regardless of the number of years remaining, whether 1,000, 2,000, or 10,000, to the best of our knowledge the supply is still finite.

In 1956, Geologist M. King Hubbert introduced an influential model that accurately predicted that oil production in the United States would begin a permanent decline around 1970.[120] His model, a bell curve, illustrated the rise, peaking, and decline in production over time. Another prediction, that global oil production would peak in 2000 proved premature, as did his forecasts of declining American and global natural gas production.

To delay the exhaustion of our fuel reserves, the Ehrlichs and others recommend imposing taxes on the use of carbon-based fuels. By raising their price, the higher taxes would discourage their use and encourage switching to alternatives. It is also argued that government could use the taxes to fund energy research.[121]

In the past, other societies have actually faced the exhaustion of their fuel supplies. "In an attempt to maintain supplies of wood fuel, [the ancient Egyptians] extended their empire further and further south. . . into the jungles of Africa."[122] Some historians believe that they overextended themselves, leading to a decline in their power.

Similarly, in the sixteenth century, England's forests, which had long supplied wood for fuel as well as for building materials, were being rapidly depleted. Fortunately for the English, they found a substitute—coal. But coal required more technology to find and procure than did wood.

So when will the world's supply of fossil fuels be depleted? And when it is, will we have the technology to provide adequate substitutes?

The following table shows the world's proved and probable carbon-based fuel reserves as of 2002.

---

[120]M. King Hubbert, "Nuclear Energy and Fossil Fuels," *Drilling and Production Practice* (Washington: American Petroleum Institute, 1956), pp. 16–18.

[121]Paul Ehrlich and Anne Ehrlich, *Betrayal of Science and Reason*, p. 96.

[122]Edward Cassedy and Peter Grossman, *Introduction to Energy*, p. 6.

WORLD CARBON-BASED ENERGY SUPPLIES (YEAR-END 2002 ESTIMATE IN YEARS OF CURRENT CONSUMPTION)
Comparing proved reserves and probable resources, the world supply of coal is several times greater than that of oil and gas combined (on an energy equivalent basis). *Source: See Appendix F.*

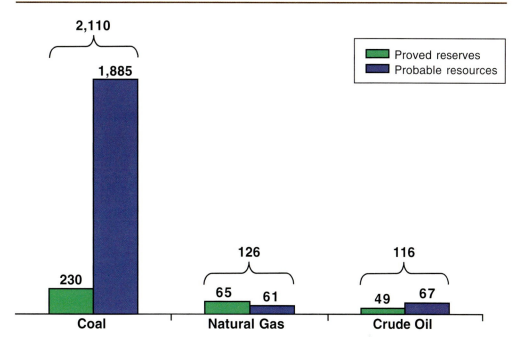

The term *proved reserves* refer to those resources that have been discovered and are currently economically recoverable, while *probable reserves* (sometimes called *resources*) include those additional amounts that can be expected to be recoverable under realistic price and technology changes. The estimates of years left in the chart are based on current consumption rates.

Hundreds of years of probable coal and hydrocarbon reserves remain at current consumption rates, though these rates will accelerate as China, India, and other poor nations industrialize. However, as people in these countries become freer, their know-how and financial capital can be expected to help make energy more plentiful and useful, not less.

Even in a worst-case scenario, resources would not disappear overnight. Instead, they gradually become harder and more expensive to find and produce. People would have time to develop the technology necessary to deal with the growing scarcity whether on the demand-side (increasing conservation) or

By the end of 1944, crude oil proved reserves were 51 billion barrels worldwide. After 58 years of production, reserves had grown to 1,266 billion barrels. In the United States over approximately the same period, 143 billion barrels of oil were produced while proved reserves increased from 20 billion to 23 billion barrels.[123] *Source:* See Appendix F.

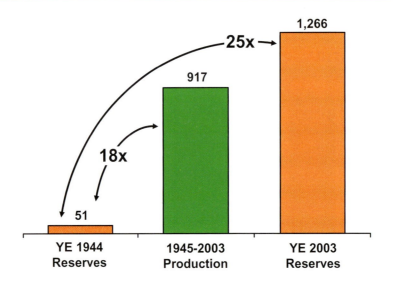

WORLD NATURAL GAS (TRILLION CUBIC FEET—PROVED RESERVES)

Even though 2,563 trillion cubic feet of natural gas were produced worldwide between 1967 and 2003, proved reserves increased six times—from 1,041 trillion cubic feet to 6,076 trillion. *Source:* See Appendix F.

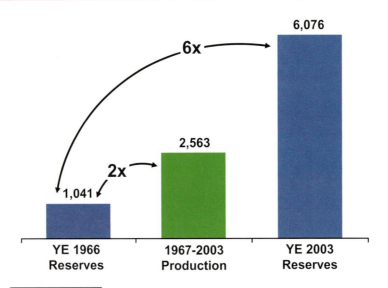

[123]Robert Bradley, Jr., *Julian Simon and the Triumph of Energy Sustainability* (Washington: American Legislative Exchange Council, 2000), p. 33.

WORLD COAL (BILLION SHORT TONS—PROVED RESERVES)
From 1950 to 2002, world proved coal reserves increased more than fourfold from 256 to 1,089 billion short tons.[124] *Source:* See Appendix F.

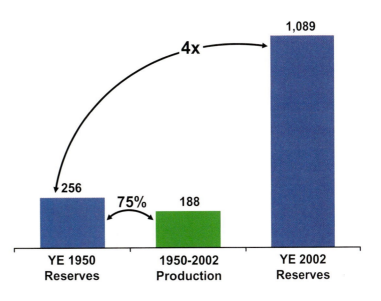

supply side (developing substitutes). Changing price signals are key to these adjustments. In fact, energy transitions over the centuries have been, with few exceptions, "remarkably orderly."[125]

For example, when an oil well is first drilled, natural reservoir pressure is usually enough to force the oil through the well all the way up to the surface.[126] Once this pressure is depleted, other means of forcing the oil up the well bore are needed. Often pumps are used to pull the oil to the surface. Natural reservoir pressure together with pumps might recover a quarter of the original oil in place (OOIP).

Next, water can be injected into surrounding wells to push the oil towards production wells. *Waterflooding* will produce another five or ten percent of the OOIP. After that, steam or CO_2 injection might be employed. Typically, little more than a third of a field's OOIP is ever recovered. "The remainder stays

---

[124]A short ton is 2,000 pounds (907.19 kg) as opposed to a long ton, which is 2,240 pounds (1016.05 kg). A metric ton is 1000 kg or 2,105 pounds. In this book, the word "ton" will mean short ton unless otherwise noted.

[125]Vaclav Smil, "Perils of Long-Range Energy Forecasting: Reflections on Looking Far Ahead," *Technological Forecasting and Economic Change*, vol. 65 (2000), p. 257. Smil refers to the historical research done by Cesare Marchetti (at fn. 37, p. 264).

[126]The terms "reservoir" and "pools of oil" are very misleading. People who hear them generally think of huge underground caverns filled with oil. In reality, there are no such caverns. Instead, the oil is contained in the pores, or "vugs," of porous rock. The more porous, or "vugular," the rock, the more readily the oil can flow towards the well bore.

PhotoDisc

behind as a potential target for new technology and/or future improvements in market conditions [i.e., higher prices]."[127]

Ultimately, oil can be mined from shallow fields. Tunnels are dug under the reservoir and holes are drilled up through the tunnel roofs into the reservoir itself. Gravity does the rest. Such mining techniques can recover more than 90 percent of a reservoir's oil. However, *oil mining* is very expensive and currently, at least, is rarely done.

Each successive recovery method is more difficult and more costly. But higher prices spur improvements in the short-to-medium term, and technological change drives down finding costs over the longer term. Offshore oil production is probably the best example of this.

As traditional crude oil reservoirs are depleted, other sources of oil will be tapped. For example, the United States has tremendous reserves of oil shale in the western United States. Globally, it is estimated that there is more than 200 times more oil in oil shale than in conventional petroleum reservoirs![128] Extraction of petroleum from oil shale is expensive, but it will become cheaper as technology progresses.

Breakthroughs in technology have made possible the refining of thick tar or bitumen. Venezuela boasts perhaps a trillion barrels of the so-called "fourth fossil fuel"—an amount equal to the world's total proved reserves of crude oil. Another one-to-two trillion barrels of heavy oil is contained in the Athabasca oil sands of northern Alberta in Canada, several hundred billion barrels of which are now categorized as proved reserves or probable resources.[129]

---

[127]Edward Porter, *Are We Running Out of Oil?*, p. 8.

[128]Bjørn Lomborg, *The Skeptical Environmentalist*, p. 128.

[129]Estimate of the Alberta Energy and Utilities Board at http//:www.eub.gov.ab.ca/bbs/products/newsletter/2003–06/feature_01.htm.

ESTIMATED GLOBAL OIL SUPPLIES

At year-2000 consumption rates, the world has many thousands of years of crude oil and crude oil substitutes (heavy oil, oil sands, and oil shale) remaining. These figures do not even take into account other carbon-based fuels such as coal and natural gas. Note that reserves in this context means proved reserves, and resources means probable future reserves. *Source:* U.S. Energy Information Administration *International Energy Outlook 2002,* p. 32, and the U.S. Energy Information Administration *International Energy Outlook 2001,* p. 47.

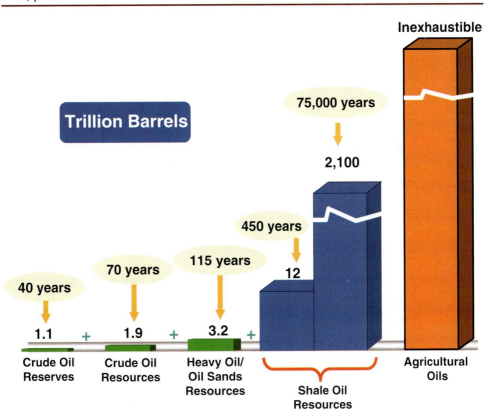

Inexhaustible

75,000 years

2,100

Trillion Barrels

450 years

12

115 years

70 years

40 years

1.1   +   1.9   +   3.2   +

| Crude Oil Reserves | Crude Oil Resources | Heavy Oil/ Oil Sands Resources | Shale Oil Resources | Agricultural Oils |

If a resource "depletes," market signals change. Higher prices check consumption, and substitutes, which were previously uneconomical, are put into service.

As fossil fuel reserves are consumed, people will switch to synthetic oil and perhaps other sources of energy that we cannot even imagine today.

❝As Sheik Yamani, Saudi Arabia's former oil minister and a founding architect of OPEC, has pointed out: 'the Stone Age came to an end not for a lack *of stones, and the oil age will end, but not for a lack of oil.' We stopped using stone because bronze and iron were superior materials, and likewise we will stop using oil, when other energy technologies provide superior benefits.* **"**[130]

Bjørn Lomborg, Danish professor of statistics

Remaining carbon-based deposits can satisfy the world's energy needs for hundreds or thousands of years. Long before that fuel is expended, technology will advance beyond anything we can possibly comprehend today. The progress in just the last century has been astounding, and the pace of change is accelerating. Worrying what people in the future will do for energy is a bit like a 19th century Arab stockpiling camel dung for the fuel needs of his 20th century descendants.

The table on the next page shows that humanity has a long tradition of overcoming shortages and leaving itself better off in the process. In fact, without the stimulus provided by need, much of the technology that we take for granted today might never have been created.

While M. King Hubbert was premature, perhaps by centuries, in his prediction that global petroleum production would peak in 2000, he got one thing right, eventually oil production will start to decline. But this will only happen when a cheaper, or otherwise better, source of energy is found.

No doubt the production of stone tools also followed a rising, then falling curve as people gradually shifted to bronze, and later to iron, implements. No doubt the graph of whale oil production followed a similar path.

Energy depletionists concentrate on current sources of energy and their inevitable decline. As a result, they see a bleak future for the world. Expansionists, by contrast, are less interested in any particular resource than in the service that it provides. Their view of the future is brighter because they choose to focus not on limited resources, but on the limitless human mind.

"*The main fuel to speed our progress is our stock of knowledge, and the brake is our lack of imagination. The ultimate resource is people—skilled, spirited, and hopeful people, who will exert their wills and imaginations for their own benefit, and so, inevitably, for the benefit of us all.***"**[131]

Julian Simon

---

[130]Bjørn Lomborg, *The Skeptical Environmentalist*, p. 120.

[131]Julian Simon, *The Ultimate Resource*, p. 348.

MAJOR RESOURCE SHIFTS IN HUMAN HISTORY

8,000 B.C.	Climatic changes and immigration in areas of the Near East make naturally occurring food scarce. As a result, people begin the shift from living as hunter-gatherers to living as farmers.[132]
1,000 B.C.	Because of a shortage of tin, the Greeks switch from bronze (an alloy of tin and copper) to iron.[133]
300 B.C.	Mayans living in the village of Kokeal (in what is now northern Belize) recycle flint tools because of a scarcity of local flint deposits.[134]
625 A.D.	Greeks switch from hull-first ship construction to frame-first construction in response to a timber shortage. The new technique not only saves large amounts of wood but time and labor as well.[135]
1550	The English start using coal as their primary fuel because a timber shortage has sharply increased wood prices.[136]
1860	The Petroleum Age begins in the United States. Though "rock oil" has been known for millennia, it has not been extensively used as a source of fuel until a scarcity of sperm whales drives up the price of whale oil.
1896	A timber shortage and rising wood prices cause the railroad industry, one of the largest timber consumers in the United States, to begin using wood more efficiently and to employ wood preservatives. In addition, the railroads find wood substitutes and begin to build bridges out of iron and concrete and rail cars out of steel.[137] Later, in the 1920s, the railroad network stops growing in the United States altogether as people switch to automobiles as their favored mode of transportation.[138]
1941	The outbreak of World War II cuts off natural rubber supplies from the United States. America begins using recycled rubber and synthetics such as neoprene (created by DuPont in 1931), Butyl (invented by Standard Oil in 1936), and Ameripol (Goodrich, 1940).[139]

Pessimism is a grim master; it carries with it the seeds of failure. New challenges should be met with excitement and confidence, confidence born not of naiveté but of a history filled with obstacles faced and overcome. Our resources are limited only by our imaginations and by the freedom, knowledge, and drive needed to turn dreams into reality.

---

[132]Charles Maurice and Charles Smithson, *The Doomsday Myth: 10,000 Years of Economic Crises* (Stanford, CA: Hoover Institution Press, 1984), pp. 113–15.

[133]Ibid., p. 105.

[134]Ibid., pp. 107–109.

[135]Ibid., p. 97.

[136]Ibid., p. 78.

[137]Ibid., pp. 54–55.

[138]J. R. McNeill, *Something New Under the Sun*, p. 310.

[139]Maurice and Smithson, *The Doomsday Myth*, pp. 40–42.

ENERGY DEPLETIONISTS

The traditional view of natural resource production, first formulated in the 1860s by English economist William Stanley Jevons and almost a century later by geologist M. King Hubbert, is that energy production will follow a bell curve—rising, peaking, and declining as reserves deplete.

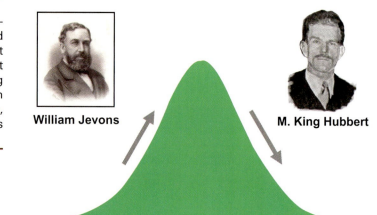

William Jevons   **M. King Hubbert**

ENERGY EXPANSIONISTS

Optimists like Erich Zimmermann and Julian Simon reject the pessimist's bell curve. Instead, they see an *energy pyramid*—expanding energy production as people use knowledge and capital to develop existing resources and discover new ones. Portrait of Erich Zimmermann courtesy of Center for American History, UT-Austin. Portrait of Julian Simon courtesy of Rita Simon.

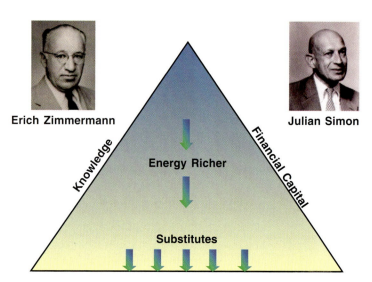

Erich Zimmermann   **Julian Simon**

Knowledge   **Energy Richer**   Financial Capital

Substitutes

People concerned about the impact of economic growth on finite resources are not the only doomsayers. Domestic oil producers and other industry allies also have promoted scary stories as part of an effort to win greater drilling incentives and more government research and development subsidies. Their message has been that without greater federal support for R&D, we could be faced with another 1970s-style energy crisis.

In 1995, an industry-led Department of Energy task force on research and development concluded, "There is growing evidence of a brewing 'R&D' crisis in the United States—the result of cutbacks and refocusing in the private-sector R&D and a reduction in federal R&D." The report warned, "The loss of our 'inventiveness'—that is, our store of human and intellectual capital—would change America's future."[140]

---

[140]Secretary of Energy Advisory Board, *Energy R&D: Shaping our Nation's Future in a Competitive World* (Washington: U.S. Department of Energy, 1995), p. 3.

Yet Daniel Yergin, the leader of the task force, testified before Congress eight years later that "a technological revolution is changing the economics and capabilities of the oil industry. . . . What technology does is lower the costs and expand the horizons. And it keeps pushing the day of shortage and depletion into the future."[141]

What did the DOE do between 1995 and 2003 to change a looming R&D crisis into a "technological revolution?" Not much. Industrial technology continued to advance through a tangle of government incentives and disincentives much as it had before.

CREATING ONE CRISIS. . .

While high resource prices can be painful in the short-term, they are really symptoms of deeper problems. They serve as warnings of shortages, and, at the same time, provide incentives for overcoming those shortages. Using price controls to solve the problem of rising prices is like trying to cure a child's fever by adjusting the thermometer.

By eliminating the feedback that free-market prices provide, price controls can quickly create shortages where none existed before or make existing shortages worse. To understand how this can happen, consider a commodity like apples. Right now there is no shortage of apples; we can go to any grocery store and buy more than we could possibly eat in a month. Suppose, though, that the president of the United States were to announce that from now on, apples will cost only a penny apiece. Within a day or two, they would disappear from the stores.

Why? Well, at one cent each, the demand for apples would skyrocket. Who could resist a bargain like that? On the other hand, at that price who could afford to grow and harvest apples, much less transport them to stores? Forcing prices below their natural market levels encourages consumption and discourages production. The inevitable result is a shortage.[142]

This sort of thing has happened repeatedly in the United States with energy production. For example, to help finance both World Wars, the federal government inflated the money supply.[143] This resulted in rapidly rising prices as the number of dollars increased faster than did the production of goods and services that the dollars could purchase. The government responded with price controls. By keeping the price of fuel below its market-clearing price, these controls encouraged consumption and discouraged production—the

---

[141] Daniel Yergin, quoted in Neela Banerjee, "Oil's Pressure Point," *New York Times*, April 3, 2003, p. 3–1.

[142] Conversely, a surplus (or oversupply) can be caused by artificially setting the price too high, as the federal government does for some farm products. High prices discourage consumption and encourage production. The result is often government warehouses bulging with unsold goods. Such overproduction means that resources were wasted (and pollution generated) to produce products that no one wants.

[143] Inflation is a general rise in prices virtually always caused by a government printing too much money or expanding credit (and with it, debt—the flip side of credit) excessively.

DYNAMICS OF INTERVENTIONISM

Government intervention into the economy often is a *cumulative process*. The most common scenario in energy has been triggered by monetary inflation (the *initial intervention*) causing prices to rise. This is followed by *subsequent intervention*: price controls to address inflation, and allocation controls to address the shortage created by price regulation.

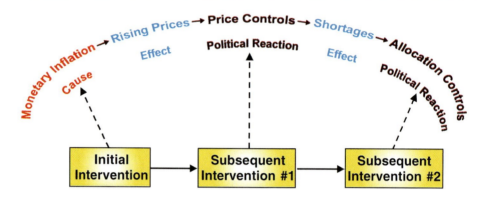

precise opposite of what was needed. In each case, a boom in oil production and a drop in prices followed post-war decontrol, thus clearly revealing the self-defeating nature of the government's intervention.

To help finance the war in Vietnam and the War on Poverty, President Lyndon Johnson again inflated America's currency. Johnson's successor, Richard Nixon, tried to treat the symptoms of inflation by imposing wage and price controls in 1971. These controls, coming during the driving season and before winter, locked in seasonally high gasoline prices and low fuel-oil prices.

As a result, refiners were encouraged to continue gasoline production at the expense of fuel-oil yields. With the coming of cold weather, heating oil supplies became tight. The federal government, along with some state governments, readied, and in some cases implemented allocation plans for fuels in short supply (butane, propane, natural gas, and fuel oil). In August 1973, the newly created Office of Energy Policy unveiled an allocation plan to be used in the event that fuel oil shortages occurred during the coming winter.[144]

On October 6, 1973, Egypt and Syria attacked Israel on Yom Kippur, a Jewish holy day. After 16 days of fighting, the war ended in a truce. Because the United States and the Netherlands had supported Israel during the brief fight, Arab oil producers declared an embargo against both countries (that is, they refused to sell them oil).

---

[144]Robert Bradley, Jr., *The Mirage of Oil Protection* (Lanham, MD: University Press of America, 1989), p. 133.

However, because of the way in which world petroleum markets work, oil exports still flowed to the two countries, and the embargo itself had little effect. What did have an impact, though, was that the Arabs also cut their oil production, dropping total world production by almost 5 percent.

The embargo lasted only about six months, and, other than a brief rise in prices, there should have been little inconvenience for American consumers. Unfortunately, President Nixon's price limits put oil companies in a bind. Even though the amount they had to pay for crude oil on the world market was going up, they could not legally pass their costs on to their customers.[145] Consumers, who would have been encouraged to conserve had the prices they paid at the pump been allowed to rise, saw no need to use less. At the same time, because oil companies were getting so little on the sale of their products, they had no incentive to produce more.

To correct this imbalance, the government revised its regulations to allow higher prices for imported oil and for "new oil" (i.e., oil produced from newly drilled domestic wells). The intention was to encourage companies to increase production and alleviate the supply problem.

As there were no physical differences between "old" and "new" oil, some companies illegally sold "old" oil as "new" so that they could charge more for it. A legal way of *gaming* the system was for two companies to sell oil back and forth to jack up the price (a practice known as "daisy chaining"). The government responded by requiring that for a sale to be legal, the oil had to be physically moved to ensure that the sale provided added value.

The government also tried to ease the shortage by rationing gasoline and dictating the amounts of fuel sent to each part of the country (shortfalls caused by price controls almost always trigger allocation controls). Some areas ended up with more gasoline than they needed, others far less. In fact, even though world supplies had dropped by less than 5 percent, some regions saw their supply levels fall as much as 25 percent below normal.

The predictable result of all this was the Energy Crisis, complete with gasoline shortages and long lines at service stations in many cities around the country.

---

[145]A few service station owners did charge higher prices *illegally*, however. Some of these owners were arrested and fined for price gouging. And yet, who was hurt by this crime? Their customers had the option of waiting in line to buy gas at the controlled price, or paying more with no wait. Some chose to exchange additional money for time. Should they have been allowed that choice? Other consumers found a way to exchange money for time legally. They hired surrogates to wait in the long gas lines for them. Clearly, though, this is a very inefficient and expensive way to get a fill-up.

> Under free market pricing, product allocation is handled automatically. Suppose, for example, gasoline demand rises more quickly in Los Angeles than in New York City, causing prices to go up in L.A. Oil companies, seeing an opportunity to increase profits, will shift gasoline to the West Coast. As supplies increase, prices will fall until the companies have no more incentive to send additional gasoline to the area.
>
> An economically rational allocation of goods and services is impossible without market prices to signal relative scarcity or abundance.

Lines at American gas stations reappeared in 1979 after the Iranian revolution triggered a second oil embargo. As M. A. Adelman stated in testimony before Congress, "The gasoline shortage was very small, perhaps 3 percent. Absent price control, there would have been a price increase, less than what actually occurred. But given price control, there had to be allocation: product by product, week by week, place by place. There was pressure on refiners to turn out more heating oil, then more gasoline, then more heating oil again. . . . Scattered shortages led to hoarding and panic buying and worse shortages yet—and those gasoline lines. No other consuming country cooked up this kind of purgatory for itself."[146]

Such shortages and misallocations were not fully resolved until 1981, by which time price controls on crude oil and various petroleum products had been lifted.

During the Iran-Iraq war, which lasted from 1980 to 1988, world production was cut by much more and for a much longer time than during either of the two oil embargoes (see the figure on the following page). By then, however, price controls had been largely removed, and there were neither fuel oil shortages nor lines at gasoline stations.

In hindsight, the confusing swirl of regulations that the government spewed out during the 1970s oil crisis gave consumers the worst of both worlds—higher prices and shortages. Like a pebble dropped in a pond, each government action rippled through the economy in ever-widening circles, yielding unforeseen consequences and creating demands for additional government intrusion.

. . . *AFTER ANOTHER*

More recently, the state of California faced its own energy crisis, caused by a series of "acts of God," coupled with price controls. In 1996, the state deregulated (or rather re-regulated) the electrical power industry in such a way that wholesale electricity prices (prices at which power companies buy electricity for resale

---

[146]M. A. Adelman, *Limiting Oil Imports*, hearing before the subcommittee on energy regulation, U.S. Senate, 96th Cong, 1st Sess. (Washington: Government Printing Office, 1980), p. 95.

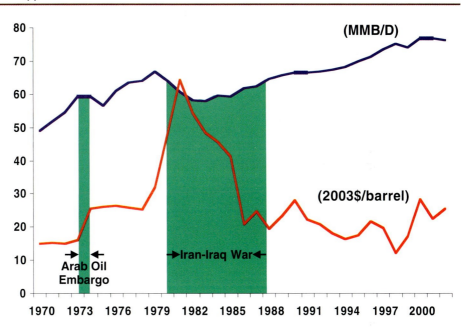

WORLD CRUDE OIL OUTPUT & INFLATION-ADJUSTED PRICES: 1970-2003

As this graph illustrates, the drop in world oil production (shown in million barrels per day) that occurred as a result of the Iran-Iraq war was deeper and longer than the drop caused by the Arab Oil Embargo. Yet the impact on American consumers was much smaller. The difference was that price and allocation controls had been lifted before the war. *Source:* See Appendix F.

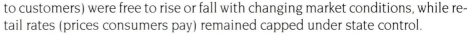

to customers) were free to rise or fall with changing market conditions, while retail rates (prices consumers pay) remained capped under state control.

Under the new regulations producers sold their electricity to a centralized state-managed power exchange at a price set by the spot market on the previous day.[147,148] Utilities purchased their power from this exchange for resale to the public.

---

[147]Under the rules, sales were made through a "reverse Dutch auction" in which all buyers pay the last price accepted during each day's trading. At the time that this scheme was established, power generation capacity was plentiful, and there was a *buyers' market* (that is, market conditions favored buyers). Under such conditions, a reverse Dutch auction tends to drive prices down. In a *sellers' market* when supplies are tight, however, this scheme drives prices up.

[148]Despite reports to the contrary, utilities were not *required* to purchase power from the Exchange at spot market prices, and *could* enter into long-term purchasing contracts with suppliers. However, any such contracts would be subject to prudence reviews whereby the California Public Utilities Commission (CPUC) could decide years after the fact that the utility had paid too much for its power. Should this happen, the utility could be required to pay back the difference between the price it paid and the price the CPUC decided it should have paid. Under such circumstances, utilities generally refused to enter into long-term contracts.

As long as plenty of power generation capacity was available, the new system worked fairly well. But in 2000, a lot of things suddenly went wrong. During that year, California had an unusually hot summer followed by an unusually cold winter. In addition, a three-year dry spell had drawn down water reservoir levels and reduced regional hydroelectric power generation by some 20 percent. Transmission line capacity problems and power plant maintenance put even more strain on the system.

On top of all this, the availability of some power plants was reduced because of environmental restrictions. During the hot summer, the plants had used up their allotted air emission allowances, and the cost of the additional allowances needed to enable their continued operation was prohibitively high.[149]

According to the California Energy Commission, 11 relatively small plants with a combined generating capacity of 1,206 megawatts came on line in California during the 1990s. Still, statewide generating capacity fell by 1.7 percent between 1990 and 1999, while demand rose by 11.3 percent during the same period.[150]

There are no coal plants in California. Nearly half of the state's electricity, and all of its peak capacity, are generated from natural gas. Therefore, the increase in demand for electric power caused by the weather, coupled with the loss of hydroelectric power, triggered a jump in demand for natural gas. While the state's four major pipelines were adequate under normal circumstances, they lacked enough spare capacity to handle all the gas that the market needed under such extreme conditions.

As a result, natural gas prices shot up and the cost of generating electricity rose with them. Unfortunately for the state's utilities, they were not allowed to pass those costs onto their customers, and they quickly went billions of dollars into debt.

In September 2000, California attempted to control rising prices by imposing a cap of $250 per megawatt-hour (MWh) for electricity sold to the state power exchange. According to Jerry Taylor, "Since wholesale prices ranged between $150–$1,000 per MWh depending on the time of day, generators responded by dramatically curtailing their sales to the California exchange."[151]

Demand quickly outstripped supply. Rolling blackouts affecting more than 675,000 homes, were used to ration power (that is, electricity was pur-

---

[149]U.S. Energy Information Administration, *International Energy Outlook* 2001 (Washington: Department of Energy, 2001), p. 127.

[150]U.S. Energy Information Administration, *Annual Energy Outlook* 2001 (Washington: Department of Energy, 2001), p. 30.

[151]Jerry Taylor, "Did Deregulation Kill California?," *Ideas on Liberty*, June 2001, p. 47.

posely cut off from one area for a few hours, then power was restored and another area cut off, and so on).[152]

Although the California Public Utilities Commission (CPUC) approved small price increases in January 2001 and larger ones a few months later, they chose to leave retail price controls in place. Instead, the state paid the difference (or at least some of the difference) between wholesale and retail prices with tax dollars. In the first half of 2001 alone, the state spent more than $8 billion for electricity.[153] So, in the end, Californians as *taxpayers* paid what Californians as *consumers* did not. They got the worst of both worlds: high prices *and* blackouts.

Struggling to find a way out, California Governor Gray Davis considered placing the power plants under state control. He and state officials even threatened to jail power-company executives who charged what the state called "excessive" prices. The most likely result of such threats is to scare off investors who might consider building power plants in California in the future. Will anyone really want to spend lots of money to build a power plant if their only reward could be the loss of their investment or even their freedom?

> **❝**Whenever there is a shortage of bread, the first thing people do is burn down the bakeries.**❞**
>
> **Ortega y Gasset, Spanish philosopher**

Although generators outside the state picked up the slack (more than 20 percent of California's power came from outside its borders), actions by the state and federal governments made it economically risky to sell power to California.

The federal government ordered power companies in neighboring states to continue to sell power to California's utility companies even though there was no guarantee that these nearly bankrupt utilities would be able to pay them back. This ruling hurt consumers in those states in two ways. First, it raised their electricity rates. Second, it increased the financial risk of running power plants in bordering states, and will probably make utilities think twice before building new facilities there.

Still, Governor Davis demanded further intervention by the federal government. Among other things, he wanted the Federal Energy Regulatory Commission (FERC) to cap power generator's profits at five percent of the cost of production. Under such a scheme, the companies could actually make more money by running up their costs and being as inefficient as possible!

---

[152]U.S. Energy Information Administration, *International Energy Outlook* 2001 (Washington: Department of Energy, 2001), p. 127.

[153]Rebecca Smith and Richard Schmitt, "Electricity Price Controls in West Are Set," *The Wall Street Journal*, June 19, 2001, p. A2.

Governor Davis also asked that price controls be imposed on the entire western region of the United States. If controls were to be effective, it was argued, they had to be applied to a large area to keep producers from diverting power away from California and toward uncontrolled states. In the face of rising political pressure to "do something," FERC gave in and on June 18, 2001, placed caps on regional wholesale energy prices. The order, which remained in effect until September 30, 2002, covered the area west of Kansas comprising almost 65 million people and nearly half of the continental United States.[154]

Within two weeks of the ruling, Las Vegas, Nevada, experienced blackouts. This was due to a clause in FERC's regulation stating that in the event of a power emergency, producers could charge Californians 10 percent more for electricity than they could consumers in other states. When July temperatures soared over 110°F in Nevada and eastern California, power producers sent their electricity where it could command the most dollars, and Las Vegas went without.

In another effort to head off rolling blackouts, the California Public Utilities Commission established the *Optional Binding Mandatory Curtailment Program*. Companies that signed up under the program agreed to cut their power consumption by as much as 15 percent within 15 minutes of a request by the CPUC. In return, the companies would be exempt from any rolling blackouts necessitated by severe power shortages. Power usage over the targeted level would be billed at $6,000 per kilowatt-hour.

The problem lay in determining the baseline from which the 15 percent reduction was calculated. It was defined as a firm's average consumption during the 10 working days before a request for power curtailment. A number of companies that signed up for the plan immediately started using as much power as they could to make it easy for them to meet their targeted reductions. In the end, the program probably made blackouts more, rather than less, likely.[155]

California's complex regulations allowed people to game the system in other ways as well. For example, suppliers were allowed to charge higher prices for imported electricity than for power generated within the state's borders. Some companies took advantage of these rules by selling electricity generated in California to out-of-state affiliates and then reselling the same power back to Californian's as "out-of-state" power.

The crisis finally ended during summer 2001 despite the fact that it was generally warmer than the summer before and more electricity was needed for air conditioning. The higher retail rates permitted by the state government, a slowing economy, and conservation programs all combined to reduce demand. In addition, new power plants went on line, and maintenance was com-

---

[154]Ibid.

[155]Joseph Menn, "Rule May Spur Firms to Waste Energy," *Los Angeles Times*, July 2, 2001, p. A1.

CALIFORNIA ELECTRICITY CRISIS: 2000-2001

California's power crisis was triggered by natural events and regulatory constraints that combined to give the state the worst of all worlds: physical shortages and wholesale price spikes. Rather than deregulate, authorities got deeper into the market by regulating wholesalers, launching new conservation programs, and signing long-term contracts.

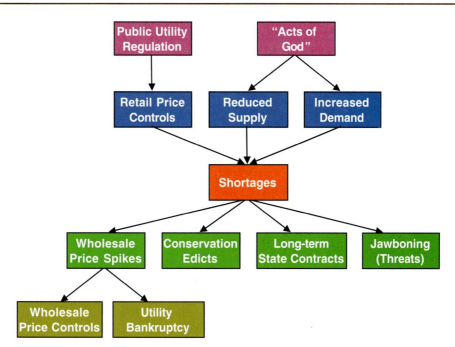

pleted on a number of older plants. Further, natural gas production increased significantly (up 4 percent in the United States and 8 percent in Canada), resulting in lower fuel prices.

In hindsight, it seems clear that the market would have ended the crisis without all the sound and fury generated by both the state and federal governments. Unfortunately, politicians wanted voters to see them doing *something* during the crisis, and that led them to create bureaucratic solutions that will no doubt get in the way for years to come.

For example in January and February 2001, at the height of the crisis, the state of California negotiated long-term (10- to 20-year) contracts with power suppliers. Less than a year later, however, the *Los Angeles Times* reported that market prices for power had "collapsed to roughly $30 per megawatt-hour— less than half of the average price in the long-term contracts."[156] In the end, consumers will pay the tab for yet another failed experiment in price control.

---

[156]Nancy Vogel, "The State: Power Contracts Improved After Freeman Left," *Los Angeles Times*, December 27, 2001, p. B8.

One of the subplots of California's power shortage was the political posturing and finger pointing that the crisis engendered. During his re-election campaign, California Governor Gray Davis worked hard to keep any blame for the debacle away from his door. In the process, he made inflammatory comments about "price-gouging" by private power companies. However, in July 2001, lawsuits forced Davis to reveal how much city-owned utilities had been charging for their power during the energy crunch. It turned out that the government-owned producers had charged an average of $344 per megawatt hour, while privately-owned companies charged less than $250.[157]

. . .AND ANOTHER

The sharp jumps in motor fuel prices experienced in the United States during 2000, 2001, and 2004 were not due to resource shortages, but rather to:

- Output quotas by OPEC. The cartel controls nearly 40 percent of the world's production and can change world crude prices overnight simply by announcing new quotas for its member nations.
- Political turmoil in Venezuela, a major oil producer.
- Unexpectedly high demand from India and China.
- Terrorist acts in Saudi Arabia (2004).
- Too little refinery capacity and too many regulations.

In spring 2001, refineries were operating at 96 percent of capacity in preparation for the summer driving months. But despite running nearly flat out, they were unable to meet demand in a manner to which Americans had become accustomed. Prices shot up.

Normally, companies seeing such high demand and high prices for their products would expand production. However, the long-term trend for refinery product prices has been downward. Temporary price spikes are not enough to justify the huge cost of constructing new facilities that will take years, not months, to complete. Although companies are expanding some existing refineries in the face of rising demand, because of thin profit margins, no new plants have been built in the United States in 25 years.

Under such conditions, with relatively few refineries all running at or near capacity, even one plant shutting down (either due to problems or for routine maintenance) can have a severe impact on the nation's fuel supplies.

Making matters worse is that city and state governments around the country have mandated that the gasoline sold locally must meet special environmental rules. As Senator Frank H. Murkowski (R-Alaska), then chairman of the Senate Energy and Natural Resources Committee and later governor of Alaska, said, "Part of the problem is that fuel made for consumption in Oregon

---

[157]Jerry Taylor and Peter VanDoren, "The Suits Tell the Tale," *National Review Online*, July 27, 2001.

U.S. REFINERIES - 2002

This map shows the locations and sizes of U.S. refineries as of 2002. Note that facilities are concentrated in coastal areas where they can receive crude oil and ship refined products by tanker. *Source:* The National Petrochemical and Refiners Association based on data from the U.S. Energy Information Administration *Petroleum Supply Annual,* vol. 1, pp. 83–97.

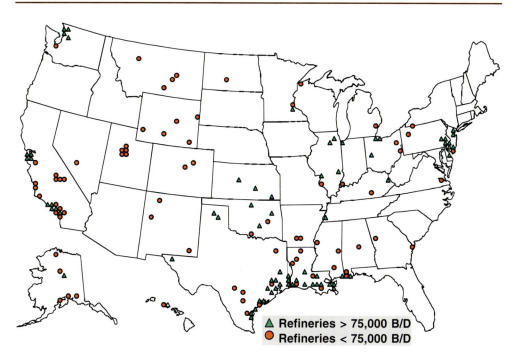

△ **Refineries > 75,000 B/D**
⬤ **Refineries < 75,000 B/D**

is not suitable for California. Fuel made for distribution in western Maryland cannot be sold in Baltimore. Areas such as Chicago and Detroit are islands in the fuel system and require special 'designer' gasolines."[158]

This confusion of laws eliminates flexibility. If a refinery producing gasoline for Chicago has to shut down for whatever reason, fuel from other areas cannot be readily shipped in to make up the difference.

Typically, when gasoline prices rise, the government's first reaction is to investigate the oil industry for possible "price gouging." Since 1973, there has been an average of about one investigation every two years. Each new investigation has begun amid ringing speeches and banner headlines, and each has cleared the industry of any wrongdoing. Often, in fact, the studies have shown that government actions were at the root of the problem. These findings usually get very little media attention.

[158]Quoted in Peter Behr, "Kicking the Gasoline 'Cocktail' Habit: Different Mixes of Grades, Ingredients in Localities Driving Costs Up," *Washington Post,* April 29, 2001, p. H1.

BOUTIQUE GASOLINES: 2001

About 20 different reformulated gasolines have been imposed on U.S. metropolitan areas not in compliance with the Clean Air Act. *Source:* American Petroleum Institute.

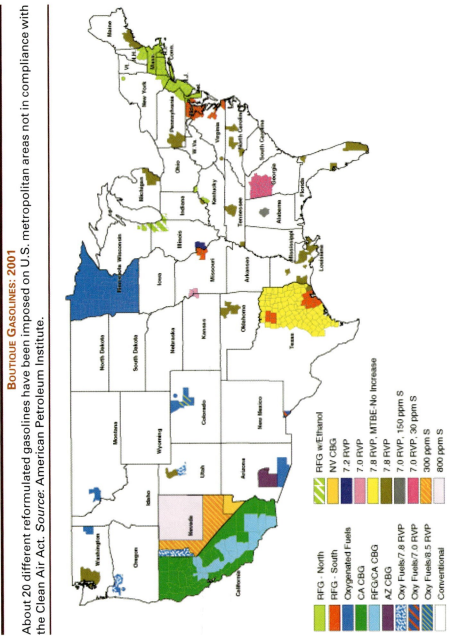

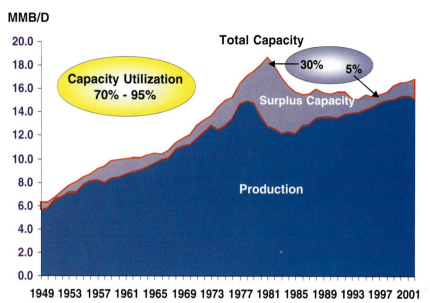

U.S. REFINERY CAPACITY & UTILIZATION: 1947-2001
The bars in this chart represent U.S. refinery capacity in millions of barrels per day, and the red line indicates the percent utilization of that capacity. While refinery capacity has increased thanks to retrofits and plant expansions, facilities are still running more than 90 percent full. *Source:* U.S. Energy Information Administration, *Annual Energy Review 2002*, p. 143.

ENERGY SECURITY

As the following chart shows, 60 percent of the United States' petroleum needs are supplied by imports, over 40 percent of which come from OPEC nations.

Many in this country are concerned about the growing dependency on foreign oil—especially on oil from nations not always friendly to the United States. In the past, government efforts to reduce imports have centered on *tariffs* (that is, taxes on imported oil), *quotas* (maximum allowable imports), or *subsidies* to domestic producers. These strategies all mean higher energy prices, higher taxes, or both. While domestic energy companies may be helped by such measures, they hurt every other industry along with every consumer.

Tariffs and quotas can anger countries that export to the United States and lead to reprisals. The most dramatic example was the creation of the *Organization of Petroleum Exporting Countries* (OPEC) in 1960 in response (at least in part) to the Mandatory Oil Import Program (MOIP), signed into law by President Eisenhower in 1959.[159] The MOIP placed quotas on the amount of oil that could be imported into the United States.

Tariffs, quotas, and subsidies generally do more harm than good, and free market economists question whether oil imports are even a problem that must be "solved" in the first place. Consider two countries, Great Britain and Japan. Thanks to North Sea production, the United Kingdom became completely self-sufficient in petroleum in the 1970s. By contrast, Japan must import all of its

[159]Robert Bradley, Jr., *The Mirage of Oil Protection*, p. 67.

U.S. PETROLEUM IMPORTS VS. DOMESTIC PRODUCTION (2003)

	Volume (1,000 B/D)	% of Total Imports	% of Total U.S. Supply
1. Canada	2,068	17	10
2. Saudi Arabia	1,772	14	9
3. Mexico	1,589	13	8
4. Venezuela	1,385	11	7
5. Nigeria	873	7	4
6. Iraq	470	4	2
7. Britain	428	3	2
8. Algeria	397	3	2
9. Angola	370	3	2
10. Norway	255	2	1
Other	2,647	22	13
Total Imports	12,254	100	61
OPEC Imports	5,175	42	26
Arab OPEC Imports	2,484	20	12
U.S. Production	7,875	–	39
Total US Supply	20,129	–	100

Source: U.S. Energy Information Administration, *Petroleum Supply Monthly,* February 2004, pp. 6, 42, 48–55, available at www.eia.doe.gov/emeu/mer/petro.html

oil. Yet each country pays about the same for petroleum as does the United States (though because of high energy taxes, consumers in both Japan and the U.K. end up paying a lot more for "petrol" than they do in the United States).

At first this may seem strange, yet consider what would happen if a barrel of oil sold for $40 in Tokyo and for $30 in London. To increase their profits, entrepreneurs would immediately start sending more petroleum to Japan. As supplies rose in Japan, the price of oil there would drop until an additional barrel would fetch no more on the market in Tokyo than in London after adjusting for transportation costs and other differentials.[160]

These same market processes make it virtually impossible for OPEC to cut off oil supplies to any single country. Suppose, for example, that Arab producers decide that they will no longer sell to the United States, and instead their oil will go only to Europe. As long as OPEC produces the same amount of oil, the effect on the United States would be negligible. Europe, now getting its fuel from the Middle East, would no longer need to purchase oil from other suppliers. These

---

[160]This example emphasizes the fact that today's oil markets truly are global. Even if the United States were to become self-sufficient in petroleum, American consumers would still pay world prices for each barrel of domestic oil.

OPEC VS NON-OPEC PRODUCTION: 1970–2003

The share of non-OPEC production has risen since the first energy crisis in 1973–74 to more than 60 percent of world production. *Source:* U.S. Energy Information Administration, *International Petroleum Monthly,* February 2004, Table 1.4.

MMB/D

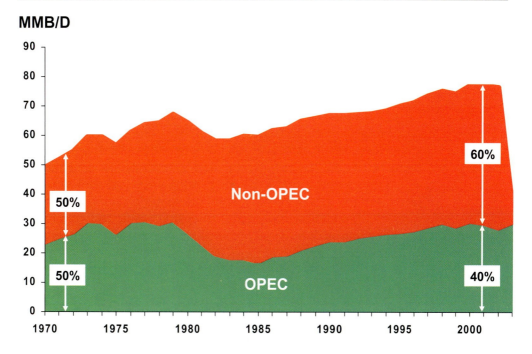

suppliers, looking for new buyers for their product, would sell to the United States. Furthermore, Europeans could turn around and resell Arab oil to Americans just as they did during the seventies oil embargo.

The only way that OPEC can really hurt the United States is by cutting back production—impacting not just America, but the world. As long as the government did not interfere with the market, the likely effect would be a temporary price hike. Higher prices will drive demand down and spur production in non-OPEC countries including the United States (assuming that it has not already used all its domestic reserves in an attempt to become "independent" of foreign oil).

Meanwhile, rising oil prices would encourage OPEC nations to cheat on their quotas and sell more of their product. In the end, the only lasting impact would be that OPEC would lose market share to other oil producing nations. As long as oil can flow freely around the world, then, the United States should not have to worry about oil imports.

Under wartime conditions oil cannot flow freely, and the United States might not be able to depend on foreign oil. A sudden loss of oil due to an outbreak of hostilities could have a serious effect on the nation's economy. On the other hand, open trade makes wars less likely because countries have little in-centive to attack their trading partners. Conversely, tariffs and trade restrictions increase the chance of armed conflict. As nineteenth century French economist Frederic Bastiat once warned, "If goods don't cross borders, armies will."

Leaders of centralized governments typically do not like open trade be-cause it makes it much more difficult for them to control their nations' economies. In a free market, for example, farmers can bypass price controls placed on food simply by exporting their produce. Similarly, consumers can avoid higher prices caused by minimum wage laws by purchasing imports rather than domestic products.

In such cases, government may respond by placing embargoes on exports and limits on imports. Historically, such actions provoked anger from nations whose people's lives or livelihoods depended upon the products or markets being restricted. They often retaliated by erecting trade barriers of their own.

As more and more trade restrictions were imposed, governments were faced with the loss of access to vital raw materials, manufactured products, or markets. In response, they acted to ensure continued access either by con-quering territory or by establishing alliances, or "spheres of influence." As a re-sult, trade wars have sometimes been followed by real wars.[161]

In the mid-1970s, the United States established the Strategic Petroleum Reserve (SPR) to store large quantities of crude oil in case of an international emergency. As many as 700 million barrels of crude oil, equal to three-to-four months of the country's total crude imports, are stored in caverns in Louisiana and Texas. The SPR is capable of pumping 3 million barrels per day (domes-tic consumption is nearly 15 million barrels a day; between 5 million and 6 million are supplied by domestic wells and the rest by foreign sources).

While the SPR does provide protection against disruptions in the oil sup-ply, it is a very expensive insurance policy. Most of the reserves were pur-chased between 1978 and 1985 when oil prices were relatively high. The average cost of acquisition in today's dollars is well over $50 per barrel. In ad-dition, the construction of the reserve and its maintenance add significantly more to the per-barrel cost of the project.[162]

Oil imports have engendered another security concern: the fear of *trade imbalances*. That is, we import more than we export and, in so doing, send money and (it is believed) jobs abroad. Though this concern is widespread, it is based on several fallacies.

---

[161]For more on this topic, refer to Brink Lindsey, *Against the Dead Hand: The Uncertain Struggle for Global Capitalism* (New York: John Wiley & Sons, 2001).

[162]U.S. Energy Information Administration, *Annual Energy Review* 2002, p. 59, and Robert Bradley, Jr., *Julian Simon and the Triumph of Energy Sustainability*, p. 122.

U.S. ENERGY TRADE: 1949-2003

The United States produces more than enough coal to meet domestic demand, but imports about 15 percent of its natural gas and around 60 percent of its crude oil. *Source:* U.S. Energy Information Administration, *Annual Energy Review 2002,* p. 11.

Quads

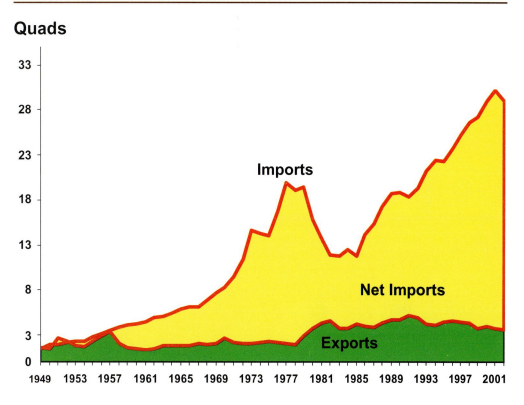

Perhaps the most glaring of these misconceptions is the belief that money is wealth. If money truly were wealth, then any nation could quickly grow rich simply by firing up the printing presses. Money is a convenient medium of exchange that lets us compare the relative value of apples and oranges, but it is the apples and oranges themselves and not the money that constitute real wealth. When OPEC nations trade oil for dollars it is not because they appreciate the engravers' art or because they have a thing for paper. They want what those dollars will buy.

Confusing money and wealth leads to another myth—the idea that a nation can become rich by exporting goods and importing money.[163] If this were

---

[163]This mode of thought is called "mercantilism" and, though it still persists, has largely been rejected by modern economists. The great danger in this way of thinking is that it tends to lead to international conflict as not all countries can export more than they import. Under mercantilism, there must be winners and losers. Open trade lets all participants win. They each trade that which they value less for that which they value more. Were this not true, no exchange would occur.

true, then why bother to build automobiles, bicycles, and computers and ship them overseas in exchange for little paper rectangles printed in foreign lands? Why not just save everyone a lot of trouble and print those pieces of paper on this side of the ocean?

When we engage in trade, we are not after foreign pieces of paper, we want foreign products. The reason we export goods is to exchange them for imports that we want more. Similarly, when people in other countries trade with us, they want our products, not our dollars.

While it is true that at any given time the country probably has "trade imbalances" with particular countries, this fact should be of no more concern than the fact that the average person has trade imbalances with the local supermarket and the gas station down the street. Dollars sent abroad can change hands many times before they return to the United States. And they may come back in forms (such as stock market investments) that are not considered when the nation's trade balance is calculated.[164]

Another aspect of energy security is the issue of whether the country will have sufficient electrical power generation and transmission capacity in the future. America's buildings, cities, transportation systems, water and sewage treatment plants, food supply infrastructure, and communications systems all depend upon a reliable supply of electrical power.

A blackout in the northeastern United States and southeastern Canada on the afternoon of August 14, 2003, highlighted just how dependent we are on electricity. This cascading system failure affected an estimated 50 million people, making it even bigger than the "Black Tuesday" outage in 1965 that left 30 million people in much the same region without power.

The blackout brought everything to a screeching halt. Subways shut down, leaving people stranded in dark tunnels far from the nearest station. Elevators stopped. The upper floors of high rise buildings, dependent upon electric pumps for water, were left without. Steel and glass office buildings with sealed windows became intolerable when their air conditioning systems shut down. Traffic became snarled when stop lights went out, and the streets were clogged with pedestrians who—left without mass transit systems—had to rely on their feet to get home. Almost 500 million gallons of raw sewage poured into New York City's waterways, fouling beaches and creating health and environmental hazards. Pumps at gas stations stopped running. Cash registers and credit cards no longer worked. Untold amounts of food spoiled in refrigerators and freezers rendered useless by the outage.

---

[164]For more information on this topic refer to Henry Hazlitt, *Economics in One Lesson* (New York: Crown Publishing, 1946, 1979), pp. 85–89.

TWO BLACKOUTS: 1965 & 2003

In American history, two blackouts have eclipsed all others: the 13-hour blackout of November 9–10, 1965, and the two-day outage of mid-August 2003. Each darkened the media center of the world, New York City (the home of *Time* magazine). Reforms implemented after the 1965 outage were intended to prevent a recurrence. *Source:* Getty Images.

Days later, it was determined that the most likely cause of the failure were downed transmission lines in Ohio. When the local transmission system shut down, the load increased on other parts of the system as they tried to meet the demand. As these systems heated up, failsafe devices shut them down before they could be damaged. Like a wave of falling dominoes, one part of the transmission grid after another went down and the lights went out.

The crisis brought out the best in average citizens and the worst in politicians. People pitched in to help each other, and utility workers labored heroically to bring the grid back up. Meanwhile the nation's politicians pointed fingers and ducked blame as fast as they could—long before the cause of the blackout was even known.

Special-interest groups tried to use the blackout as evidence to support their own particular views: deregulation, re-regulation, alternative energy, nuclear power, hydroelectric power, government investment, private investment, and on and on. Some pundits argued that the blackout proved that utility deregulation had failed. Supporters of deregulation countered that the failed transmission system was fully regulated, and that the low profits allowed by the regulators left no incentive for companies to expand and improve the grid. Opponents fired back that it was the deregulated power generation system that overloaded the grid in the first place. They argued that only complete regulation of both the generation facilities and the transmission lines could ensure the necessary coordination between the two.

We offer two observations and a simple question: The (partially) deregulated portion of the industry, power generation, kept up with consumer demand. The regulated portion of the industry, transmission, did not. Which, then, is the better way to go—toward more regulation, or less?

ENERGY AND THE ENVIRONMENT

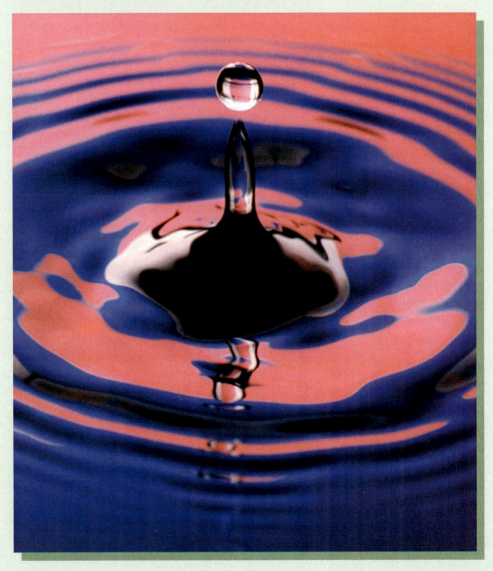

5

DEFINING POLLUTION

Every action of every living thing uses resources and produces pollution. Even while you are just sitting and reading this book, you are consuming oxygen (a resource) and producing carbon dioxide (a pollutant). Similarly, trees consume carbon dioxide (a resource) and produce oxygen (a pollutant).

Wait! First we called oxygen a resource and carbon dioxide a pollutant, and then we turned around and called carbon dioxide a resource and oxygen a pollutant! That doesn't make sense, does it? Well, it all depends on which side of the fence you are sitting. If you happen to be a tree, oxygen is something you are trying to get rid of (remember, trees give off oxygen), and carbon dioxide is something that you need to survive. If you are a human, on the other hand, you need oxygen to breathe, and you have to get rid of carbon dioxide. Nature has a way of balancing things out. We need trees, and trees need us.

So when is a substance a resource, and when is it a pollutant?[165] The question is best answered by example. There are a number of natural *oil seeps* in the floor of the Pacific Ocean off the coast of California. The amount of oil flowing into the water is small and poses no danger to sea life in the area. In fact, given that petroleum is an organic substance (that is, it is carbon-based), it is biodegradable and serves as food for microbes, which are, in turn, eaten by larger organisms, and so on up the food chain. In small amounts, then, crude oil actually acts as a fertilizer.

---

[165]Currently, carbon dioxide generated by human activity is called an "emission," while carbon monoxide is designated as a "pollutant." The distinction is important for legal reasons because the U.S. Environmental Protection Agency has the authority to regulate substances designated as pollutants. There is ongoing debate about how to determine the point at which an "emission" becomes a "pollutant."

However, suppose that an oil tanker were to run aground and spill millions of gallons of crude into the water. That amount of oil would be so overwhelming that it might take years for microbes to break it all down. In the meantime, it would almost certainly kill thousands of fish, birds, sea mammals, and other creatures. In this case, the oil is clearly a pollutant.

Similarly, while we think of sewage as pollution, a small amount of it in a river does no harm and maybe even a little good for some passing bacteria. But dump an entire city's waste into a river day after day, and the river will quickly become a stinking cesspool devoid of any life larger than a germ.

Finally, consider a piece of radioactive material. Though small, the material may emit radiation that can harm living things many feet away. Worse, the material may remain dangerous for tens of thousands of years. Even a small amount of such material could be considered a pollutant. As with other types of pollution, however, the danger is in the dose. A little radiation exposure is safe; a lot can be harmful or even fatal.

In sum, if the substance in question decays fairly rapidly and provides benefit to some living creature, it's probably a "resource." If it lingers on and especially if it harms or destroys life, it is a "pollutant."

Even when toxic substances are involved, the most important factor in determining whether something is a serious pollutant is quantity and nature's ability to deal with that quantity. Smoke from a few campfires is of little concern. Wind will disperse the smoke, and the next rain will clean any remaining particulates from the air. However, concentrate thousands of people in a city—all burning wood, peat, or coal to cook their meals and keep warm—and the sky turns black. Fill the local river with the city's sewage and refuse, and you have Shakespeare's London of sixteenth century England. Pollution is not a new problem.

So what is the answer? Should we turn back the clock and live as Stone Age peoples did? Anthropologists are beginning to suspect that that way of life was not as environmentally friendly as previously believed. The extinction of a number of species around the world including "the moas of New Zealand, the giant lemurs of Madagascar, and the big flightless geese of Hawaii" coincided with the appearance of humans.[166]

In the Americas, early hunters are probably responsible for the demise of mammoths, the Shasta ground sloth, and Harrington's mountain goat.[167] American Indians often hunted in wasteful and destructive ways. One of their techniques was to lead whole herds of buffalo over a cliff.[168] Another common tactic, known as "box burning," was to set fires all around a herd.

---

[166]Jared Diamond, *Guns, Germs, and Steel*, p. 43.

[167]Ibid., p. 47.

[168]Head-Smashed-In Buffalo Jump, near Calgary in Alberta, Canada, is one of the best-known and well-preserved buffalo jumps. Aboriginal peoples used the site continuously for more than 5,500 years.

Such a way of life is so unproductive and wasteful that it could only support a fraction of the people now living in the world today. That might not be so bad if you are one of the few chosen to live, but even then it would be no picnic. Typically, Indians led relatively short, disease-ridden lives. Tribal members too old or sick to pull their own weight were often, quite literally, left for the wolves. All in all, there is a lot to be said for indoor plumbing, painless dentistry, and retirement plans.

If the answer is not for us all to go back to living in buffalo skin tepees or mud huts, what *do* we do? A growing number of economists believe that the answer lies in the efficiency and inventiveness unique to people living in free societies.

INEFFICIENCY, WASTE, AND POLLUTION

Inefficiency is waste, and waste is pollution. For example, there is waste when fuel does not burn completely (i.e., when it burns inefficiently). The unburned portion of the fuel either goes up the chimney or must be hauled away to a dumpsite—pollution.

Before wood can be used as a fuel, it must first be hauled to the site where it will be used (this includes hauling the part of the wood that will not be burned as well as the part that will). Transportation costs resources (fuel) and produces pollution (engine or animal emissions). When the wood is burned, soot, smoke, and ashes (unburned materials) either go up the chimney (pollution) or must be carted away (more transportation costs and more pollution).

Natural gas, on the other hand, is a very efficient fuel. It burns almost completely so that little energy is expended or pollution created in either transporting useless material to the power plant or in hauling unburned ashes away. In addition, far fewer emissions go up the chimney.

So should people be forced to act more efficiently? Fortunately, free markets automatically provide incentives. In a free market, people are encouraged to act efficiently in order to save money. In doing so, they usually end up saving resources thereby reducing waste and pollution.

In the book, On *the Economy of Machinery and Manufactures*, written more than 170 years ago, Charles Babbage, inventor of the first mechanical computer, observed, "amongst the causes which tend to the cheap production of any article, and which are connected with the employment of additional capital, may be mentioned, the care which is taken to prevent the absolute waste of any part of the raw material."[169]

In 1862, journalist Peter Simmonds explained how the waste from woolen mills became a source of profit. "By means of mechanical appliances and chemical action, the refuse formerly turned into the river Nith to the injury of

---

[169]Charles Babbage, On *the Economy of Machinery and Manufactures* (London: Charles Knight, 1832).

the salmon, is made to produce stearine, which forms the basis of composite candles, as well as a cake manure that sells at 40s [shillings] per ton."[170]

In the early nineteenth century, coal gas (methane plus some amount of impurities) was used as an illuminant (i.e., a fuel burned to provide light) in parts of England, but it was not popular because of the unpleasant smells that were produced when it was burned. Chemists learned to purify coal gas and remove the noxious substances, however. These substances became profitable by-products of the coal gasification process. As an observer of the time put it, "the waste and badly-smelling products of gas-making appeared almost too bad and foetid for utilization, and yet every one of them, Chemistry, in its thriftiness, has made almost indispensable to human progress."[171]

Or consider the early days of the petroleum refining industry. Petroleum was originally valued chiefly because it could be refined to produce kerosene, which was used as an illuminant. Naphtha was an unwanted by-product of the refining process. Most refiners either burned it or simply let it evaporate. Naphtha, a low-grade illuminant, could occasionally be sold at a profit, however. William Rockefeller, a partner with his brother, John D. Rockefeller, found that he could increase his company's profits by storing naphtha in tanks when prices were low for later sale when prices recovered.[172]

These early businessmen probably had no intention of protecting the environment. Yet their desire to reduce costs and increase their profits led them to take actions that did exactly that.

This market-driven search for profits has, over time, moved people in western nations to reduce waste and use resources ever more efficiently. As a result, the air and water in these countries have been getting progressively cleaner even as population, production, and fuel combustion have increased.

OUR IMPROVING ENVIRONMENT

Within just a few decades, market incentives and improving technology combined with laws and regulations have had a dramatic effect on our country's environment. Between 1970 and 2002, emissions of the six so-called "criteria air pollutants" in the United States dropped anywhere between 17 percent (nitrogen oxides) and 98 percent (lead).[173]

---

[170]Quoted by Pierre Desrochers, "Saving the Environment for a Profit, Victorian-Style," *Ideas on Liberty*, May 2003, p. 32.

[171]Ibid., 34.

[172]David Hawke, *John D.: The Founding Father of the Rockefellers* (New York: Harper & Row, 1980), p. 55.

[173]U.S. Environmental Protection Agency, *Latest Findings on National Air Quality: 2002 Status and Trends* (Washington: EPA, 2003), p. 2.

U.S. AIR EMISSIONS: 1970 VS 2002

Data compiled by U.S. Environmental Protection Agency indicate a significant decline in all six of the criteria air pollutants in the past three decades. *Source:* U.S. Environmental Protection Agency, *National Air Quality: 2002 Status and Trends,* p. 2.

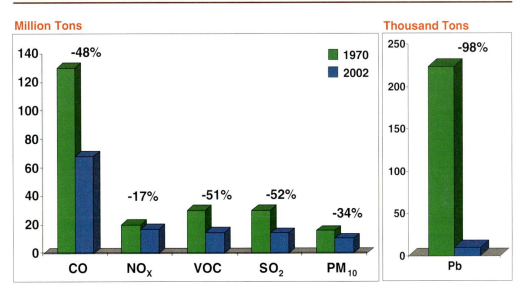

In aggregate, air pollution fell by nearly half despite significant increases in population, gross domestic product (GDP), vehicle-miles driven, and energy usage.[174]

U.S. GROWTH VS. AIR POLLUTANT EMISSIONS

Air pollution in the United States dropped significantly while, at the same time, the country was growing in both population and wealth. *Source:* U.S. Environmental Protection Agency, *National Air Quality: 2002 Status and Trends,* p. 4.

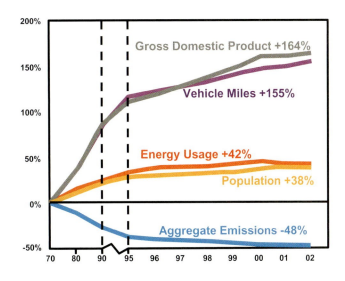

[174]Ibid., 2.

For decades, Pittsburgh was one of the most polluted cities in America. Between 1930 and 1950 the city experienced about 350 smoky days each year! However, by 1970 (the year the Clean Air Act was passed), that number had dropped to 200 days out of the year, and by 1990 to fewer than 20.[175]

Smog in Los Angeles and Houston is currently considered to be the worst among cities in the United States, yet the air quality in both cities has rapidly improved. Over the last two decades, the number of days in which the ozone levels exceed limits set by the U.S. Environmental Protection Agency (EPA) has been steadily dropping.

LOS ANGELES VS. HOUSTON OZONE VIOLATION DAYS

Ozone (smog) violation days have decreased by almost three-fourths in Los Angeles and one-fourth in Houston since the early 1980s. Houston recorded more "episode days" than Los Angeles for the first time in 1999–2000. *Source:* See Appendix F.

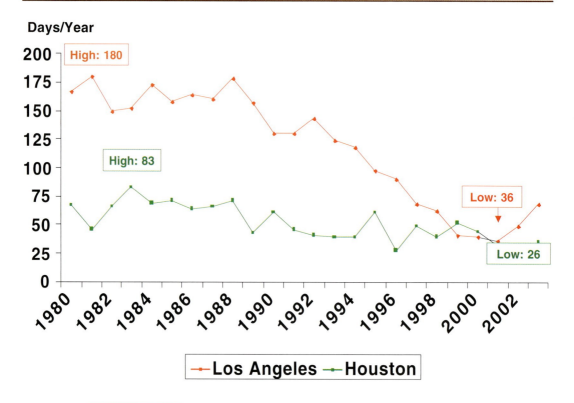

Days/Year

High: 180
High: 83
Low: 36
Low: 26

— **Los Angeles** — **Houston**

---

[175]Cliff Davidson, "Air Pollution in Pittsburgh: A Historical Perspective," *Journal of Air Pollution Control Association* 29 (1979), pp. 1035–41 and various issues of *Council on Environmental Quality Annual Report.*

U.S. Motor Vehicle Emission Reductions

EPA emission standards for volatile organic compounds (VOC) and nitrogen oxides (NO_x)—the precursors to smog—are expected to reduce concentrations of these gases by 99 percent as compared to levels in the mid-1960s. *Source:* U.S. Environmental Protection Agency; Alliance of Automobile Manufacturers.

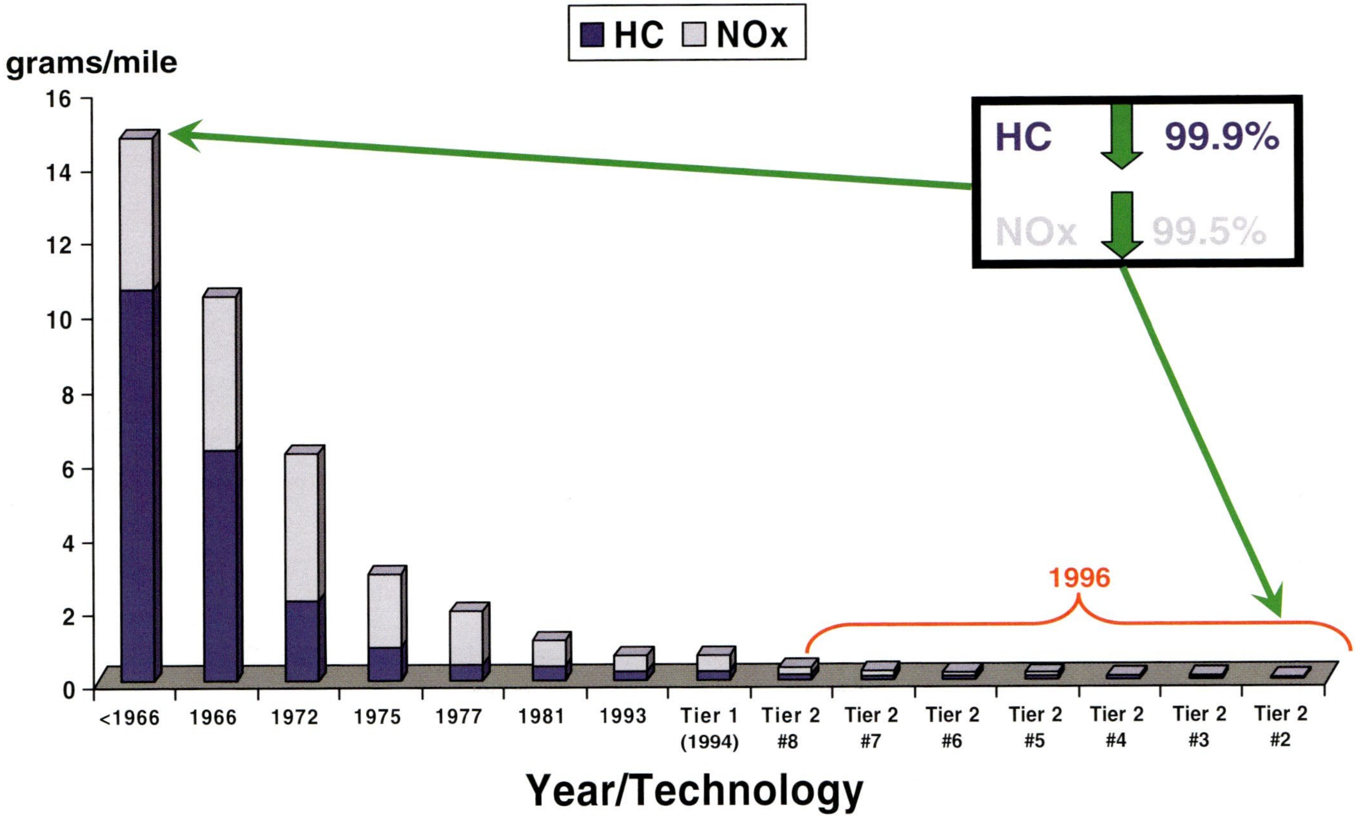

HC 99.9%

NOx 99.5%

■ HC ☐ NOx

grams/mile

16
14
12
10
8
6
4
2
0

<1966   1966   1972   1975   1977   1981   1993   Tier 1 (1994)   Tier 2 #8   Tier 2 #7   Tier 2 #6   Tier 2 #5   Tier 2 #4   Tier 2 #3   Tier 2 #2

1996

Year/Technology

Corbis

The Clean Water Act of 1972 gave the federal government the power to set and enforce national water quality standards and to regulate the dumping of industrial and municipal wastes. Within 25 years most easily identifiable sources of water pollution were brought under control.[176] According to Stephen Moore and Julian Simon, "by 1994, 86 percent of U.S. rivers and streams were usable for fishing and swimming—up from 36 percent in 1972."[177]

The number and size of oil spills in the United States have decreased significantly since 1990, in part because major oil companies have replaced single-hull oil tankers with double-hull ships since the Valdez oil spill.

Over the same period, farms have become far more productive so less land is needed for agriculture. Because of this, and because less wood is being burned for fuel, our nation's forests are expanding.

In other parts of the developed world, the environment has been getting better as well. London's air pollution peaked around 1890 and has been dropping ever since. In fact, the city's air is cleaner than it has been since the late 1500s, and the famous London fogs are becoming things of the past.[178] The Thames River, which had been without fish for a century, by 1968 boasted some 40 varieties.

Some former Eastern-bloc countries such as Poland and the Czech Republic have also seen improvements. As Andrew Steer with the World Bank noted, eastern European lead smelters have cut emissions to about one-sixtieth of their previous levels "as a result of improved housekeeping and modest investments."[179]

---

[176]Paul Ehrlich and Anne Ehrlich, *Betrayal of Science and Reason*, p. 52.

[177]Stephen Moore and Julian Simon, *It's Getting Better All the Time* (Washington: Cato Institute, 2000), p. 188.

[178]Bjørn Lomborg, "The Truth About the Environment," *The Economist*, August 4, 2001, p. 64.

[179]Andrew Steer, *Ten Principles of the New Environmentalism* (Washington: The World Bank, 1996), p. 6, available at http://www.worldbank.org/fandd/english/1296/articles/0111296.htm

---

OIL SPILLS IN U.S. WATERS (MILLIONS OF GALLONS PER YEAR)

Oil-spill volumes have significantly decreased, especially after industry and regulatory reforms were made in response to the 1989 Valdez spill. *Source:* U.S. Coast Guard, *Annual Data and Graphics.*

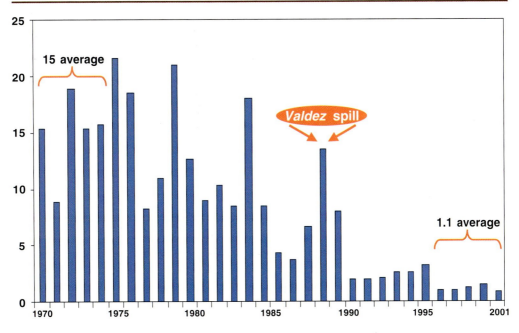

SOME COMPARISONS

The environmental picture is not nearly as bright in other parts of the world. In 1996, Paul and Anne Ehrlich observed, "Huge cities such as Mexico City, Sao Paulo, Jakarta, Bangkok, Beijing, Delhi, and Nairobi have horrific smog despite being located in countries with far less industry than the United States. Tens of thousands to millions of cars, trucks, and buses with no smog controls cram the streets; hundreds of uncontrolled factories, smelters, and power stations belch smoke and pollutants; and in some cities millions of open cooking fires foul the air. Third-world rivers are often essentially open sewers spiked with pesticide cocktails."[180]

In some of the countries that made up the former Soviet Union, pollution-control laws are ignored and little attention is paid to energy efficiency. As a result, the environment in these countries is in such terrible shape that it has significantly damaged the health of the people who live there. The Ehrlichs pointed out that, "During the 1970s, mortality rates in the Soviet Union stopped

---

[180]Paul Ehrlich and Anne Ehrlich, *Betrayal of Science and Reason*, p. 59.

falling and began rising, and the government, presumably embarrassed, stopped publishing mortality statistics. The trend worsened after the Soviet Union's breakup."[181]

Other than global warming (discussed in the next chapter), the main environmental issues that the world faces center on the Third World. Air and water pollution do not respect international borders—dirty air created in one country can quickly become another's problem. How can these nations move toward cleaner, more efficient fuel sources? Can they make this switch and achieve their own Industrial Revolution without reproducing the same environmental problems created during the West's revolution?

Before these questions can be addressed, another more basic question must be answered. *Why is there such a difference between the environments in these countries and those of the countries in the industrialized West?*

Some believe that the problem is a lack of proper environmental laws in the Third World. Yet the Soviet Union had strong laws on the books—they were simply ignored.

The difference is poverty. Third World countries are much poorer than western nations. When people are worried about where their next meal is coming from, they are much less concerned with such things as clean air and water.

Clean and efficient technology is generally more expensive than dirty, inefficient technology. No high-tech equipment is needed to burn wood for heat. But it takes a lot of costly machinery and know-how to locate, produce, transport, and use natural gas as a fuel.

> **❝**[The] *dirtiest water and air are not found in the rich countries, rather they are found in the developing nations. As pollution is rapidly becoming a global issue, worldwide prosperity should be viewed as the solution to, not the cause of the problem.* **❞**[182]
>
> *- Ziock, Lackner, and Harrison*

The next question, then, is why are these countries poor? Some suggest that it has to do with natural resources. America has plentiful resources; therefore, America is rich. But Russia also has huge resources, as do Africa, Mexico, and South America, and yet these areas are poor. At the same time, wealthy nations such as Japan, Taiwan, and Switzerland have almost no natural resources.

---

[181]Ibid., p. 60.

[182]Han-Joachim Ziock, Klaus Lackner, and Douglas Harrison, "Zero Emission Coal," *Energy* 2000: *The Beginning of a New Millennium* (Lancaster, PA: Technomic Publishing, 2000), p. 1274.

Others point to overpopulation as the problem; India and China have high population densities, so they are poor. Yet the Netherlands, Japan, Hong Kong, Belgium, South Korea, Taiwan, and Great Britain all have much higher population densities than either of these countries, and they are wealthy. At the same time, extremely impoverished nations like Ethiopia have very low population densities.

Then there are those who claim that some are poor precisely because others are rich. They believe that the world is a zero-sum game in which a few can win only if others lose. But this view can be proven wrong simply by looking around. Wealth surrounds us. Not dollar bills, but real wealth: books, computers, buildings, cars, supermarkets, houses, and factories. These are evidence that wealth can be created, and created without limit, not just redistributed. When people first appeared on Earth they had nothing. If it were true that one can gain only what another loses, we would still have nothing.[183]

While it is true that Americans consume more resources per capita than do people in other countries, we also produce more than anyone else, and there are many around the world who live off of our surplus. Moore and Simon point out that "American workers are the most productive in the world. Most industrialized nations of Europe, for example, still only have productivity rates of about 80 percent of levels in the United States. The workers in Asian nations

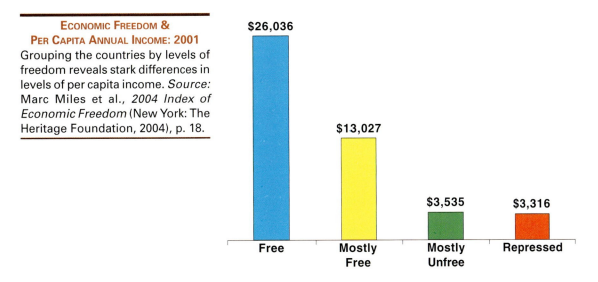

**ECONOMIC FREEDOM &
PER CAPITA ANNUAL INCOME: 2001**
Grouping the countries by levels of freedom reveals stark differences in levels of per capita income. *Source: Marc Miles et al., 2004 Index of Economic Freedom* (New York: The Heritage Foundation, 2004), p. 18.

$26,036 — Free
$13,027 — Mostly Free
$3,535 — Mostly Unfree
$3,316 — Repressed

---

[183]It is easy for individuals in government to fall into zero-sum thinking because that's how their world operates—every dollar that government spends must first be taken or borrowed from a private citizen or company. Politics is concerned with dividing the pie (that is, distributing tax dollars), while industry is concerned with making the pie bigger.

have less than 60 percent the productivity rate of American workers,"[184] while in 1999 Russian labor productivity averaged only 18 percent of U.S. levels.[185]

What wealthy nations have in common is *liberty*—the right of individuals to act as they choose without interference, so long as they don't interfere with the rights of others to do the same. Each year, the Heritage Foundation publishes its Index of Economic Freedom (*Appendix* B). The Foundation ranks the world's nations by a number of economic variables to determine which are the freest. The average person living in a repressed economy lives in poverty on an income equivalent to about $3,300 a year. People living in the world's free economies enjoy an average per-capita income of $26,000. This should not be surprising. In free societies, where life, liberty and property are protected, people have much greater incentive to create wealth, because such protections ensure that people control and benefit from the fruits of their own labors.

The problems in the Third World come not from a lack of governmental regulations, but from a lack of freedom to create, own, trade, and sell property.

FREEDOM, INCOME, & LOCATION

This scatter diagram shows that annual income rises with increasing freedom regardless of location. Freedom works in Asia and Africa as well as it does in Europe and North America. *Source:* Marc Miles et al., *2004 Index of Economic Freedom* (New York: The Heritage Foundation, 2004), p. 18.

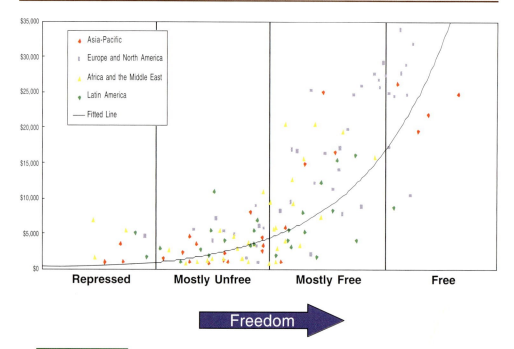

[184]Stephen Moore and Julian Simon, *It's Getting Better All the Time*, p. 96.

[185]Brink Lindsey, *Against the Dead Hand*, p. 123.

> ❝[Keys to sustainable development] "laid out in the Brundtland Commission Report in 1987 and in the Rio de Janeiro Earth Summit's Agenda 21 in 1992 [include]:
>
> - . . . a crucial and potentially positive link between economic development and the environment. . .
> - Addressing environmental problems requires that poverty be reduced.❞[186]

Andrew Steer

RETHINKING THE ROLE OF GOVERNMENT

Laws and regulations, however well meant, often make things worse. In the early nineteenth century, our government and courts decided that damage to an individual's property caused by a factory's air or water pollution was outweighed by the factory's benefit to the general public.[187] Therefore, individuals were denied the right to sue for damages caused by such pollution. However, if private property rights had been enforced, factory owners would have been required to reimburse property owners, and would have had a strong incentive to clean up their plants' waste. In the long run, this would have benefited everyone.[188]

A more recent example is the mandated use of oxygenates in gasoline to promote more complete combustion and reduce emissions. There are two oxygenates available: a chemical called MTBE (methyl tertiary-butyl ether), and ethanol. Unfortunately, both of these chemicals have serious side effects. MTBE is a suspected carcinogen, and it has been seeping into the groundwater. Traces of MTBE have been found in the groundwater of 49 of the 50 states, and a number of water wells in California have had to be shut in because of contamination.

Ethanol, on the other hand, costs more resources (and energy) to produce than MTBE and, as a consequence, is much more expensive. It also makes gasoline more volatile, causing more of it to evaporate, which adds to air pollution.

Moreover, any moisture in pipelines and storage tanks will cause ethanol to separate from gasoline.[189] As a consequence, gasoline and ethanol are shipped separately and are mixed at distribution terminals before being

---

[186]Steer, Andrew, *Ten Principles of the New Environmentalism*, p. 1.

[187]Morton Horwitz, *The Transformation of American Law, 1790–1860* (Cambridge, MA: Harvard University Press, 1977), pp. 74–101.

[188]Robert Bradley, Jr., *Oil, Gas and Government: The U.S. Experience*, pp. 1268–69.

[189]U.S. Energy Information Administration, *Annual Energy Outlook 2001* (Washington: Department of Energy, 2001), p. 36.

loaded onto trucks for delivery to gas stations. This additional handling increases the relative cost of using ethanol even more.

Finally, the use of either ethanol or MTBE increases NO_x emissions and thus helps to produce smog.

These types of unfortunate and unforeseen effects of well-intentioned regulations are all too common.

ARE REGULATIONS NECESSARY?

Some free marketers advocate doing away with regulations altogether. They argue that market incentives and property rights enforcement are sufficient to protect the environment. Certainly people who own a piece of land have more interest in maintaining its value than anyone else. They do not want their property to lose its value by being stripped of trees, fouled by factory effluent, or clouded with smog.[190] If, it is argued, the government would simply allow people to protect their property through the courts, our environment would be much cleaner.

Strengthening and expanding property laws is an important step in the right direction. Under such a system, if a company's factory polluted the air above someone's land, thereby reducing its value or harming its residents, the landowner could sue the company and receive damages. This system would achieve two important things:

1. *Justice.* The landowner would be compensated for the harm done to him by the factory.
2. *Environmental protection.* The company would be penalized for polluting, and, therefore would have a strong incentive to reduce emissions in the future.

This would work well in cases where the source of the pollution is obvious—in this example, a factory. But what if the air pollution that is damaging the landowner's property is caused by hundreds of cars that drive by every day? Whom does he take to court?

Another sort of problem occurs in situations in which no one owns the property or resource being damaged. Such cases are referred to by the phrase, "the tragedy of the commons."[191] When no one owns an object, no one has any

---

[190]People's ability to leave property to their children gives them an incentive to work to maintain their property's value beyond their own lifetimes.

[191]Garrett Hardin, "The Tragedy of the Commons," *Science*, December 13, 1968, pp. 1243–48.

incentive to maintain it. In fact, the incentive is to strip away anything of value before someone else gets it first.[192]

> **"**What is common to many is taken least care of; for all men have greater regard for what is their own than for what they possess in common with others.**"**[193]
>
> **Aristotle**

In such cases, government regulation may be the only way to protect the environment. The challenge is to create regulations that do more good than harm.

SETTING MEANS . . .

Often regulators specify the means rather than the ends. That is, instead of establishing goals (e.g., clean air or clean water), local, state, and federal government agencies may write laws and regulations that either ban or require certain methods, technologies, or materials. This means-setting, *command-and-control* approach creates a number of problems:

1. *Laws and regulations may institutionalize the tragedy of the commons.*[194] As discussed in an earlier chapter, the *rule of capture* and related regulation led companies to drill as many wells as possible in order to pump oil out of the ground before their competitors could. By encouraging companies to drill otherwise unnecessary wells, the rule led to wasted resources and sometimes to reservoir damage.

   Groundwater in the United States is still a common property resource; because no one owns it, no one has an incentive to conserve it.[195]

2. *Special interests lobby the government to get their products or services mandated by regulation.* The use of oxygenates in gasoline is a good example. Ethanol, one of the two available oxygenates, is made from corn. Farmers who grow corn, and companies that make ethanol from it, have heavily pressured Congress to require its use.

---

[192]Terry Anderson and Donald Leal, *Free Market Environmentalism* (New York: Palgrave, 2001), pp. 12–13.

[193]Aristotle quoted by Will Durant, *The Story of Civilization Vol. II: The Life of Greece* (New York: Simon and Schuster, 1939), p. 536.

[194]Fred Smith, "Enclosing the Environmental Commons," in Ronald Bailey, ed., *Global Warming and Other Eco-Myths: How the Environmental Movement Uses False Science to Scare Us to Death* (Roseville, CA: Prima Publishing, 2002), p. 300.

[195]Ibid., p. 297.

3. *Regulations can create (or destroy) entire industries overnight*. The use of such power adds uncertainty and risk to the market. If risk reaches unacceptable levels, investors put their money elsewhere. The concentration of political power in Washington forces companies to lobby Congress and the White House for protection against its arbitrary use. Corporate lobbying, in turn, increases people's distrust of the system.

4. *Regulations are often the result of compromise*. After concessions have been made to this powerful congressman or that influential senator, the resulting law or regulation may be very different from the original proposal and have very different consequences. Politics may be "the art of the possible," but what is politically possible may be neither practical nor environmentally friendly.

   Compromise can also result in laws so vaguely worded that they can be interpreted in any number of ways. In the end, it is left up to regulatory agencies and the courts to decide what a bill actually means. Their interpretations may be very different from the original intentions of the bill's proponents.

   The Clean Air Act amendments of 1977, for example, stated that only new factories and power plants would have to meet the tighter emissions standards imposed by the Act. Existing plants would continue to be regulated under the pre-existing standards unless the old plants were "substantially modified." Unfortunately, Congress did not precisely specify what "substantially modified" meant.

   In 1998, the EPA sued the owners of a number of old plants, charging that the upgrades done over the years to these plants had cumulatively added up to "substantial modifications." The owners responded, with some justification, that the EPA had originally approved their changes, and that altering the rules after-the-fact amounted to the passage of a retroactive law, something explicitly forbidden by the U.S. Constitution.

5. *Lobbyists may support regulations as a way of hurting their competition*. Utility companies with "old source" power plants, for example, welcomed the Clean Air Act's 1977 amendments because it put potential competitors at a disadvantage by raising the cost of market entry.

   Other amendments to the Clean Air Act required power companies to reduce sulfur dioxide emissions by installing scrubbers. A less expensive way to lower emissions would have been to switch to low-sulfur coal, but eastern labor unions and coal mining companies (which produce high-sulfur coal) successfully lobbied to get the requirement for scrubbers enacted into law.[196] This resulted in a waste of resources as (otherwise unnecessary) scrubbers had to be built, installed, and powered.

In the United States during the twentieth century, government intervention in the energy market has commonly been industry driven. Firms often organized lobbying groups to obtain favorable regulation or special subsidies. Free-market economist Milton Friedman complained, "Time and again, I have castigated the oil companies for . . . seeking and getting governmental privilege."[197]

6. *Regulations can eliminate or alter feedback.* Feedback is an essential component of any activity. Imagine how dangerous the world would be for a person who had lost the ability to feel pain (as happens with certain forms of leprosy). Such a person could do serious damage to himself by continuing to walk on a badly sprained ankle, or putting his hand on a hot stove without knowing it.

Government action can create a sort of institutional leprosy by weakening or even destroying the feedback loops that make it possible for companies to know whether their activities are of any value. For instance, by taxing productive companies in order to subsidize unproductive ones, governments perpetuate the waste of resources.

7. *"Hard cases make bad law."*[198] All too often, regulations are hastily written in response to the public's demands that the government "do something" in the face of a crisis. Petroleum price controls during the 1970s are a case in point. Under the provisions of the rules, refiners could charge more for higher-octane fuels, so they were encouraged to increase the lead content to artificially boost octane ratings.

At the same time that crises lead to demands for action, they tend to increase the cost of any action. For instance, in response to the power shortage of 2000–2001, the state of California negotiated long-term contracts for the purchase of electricity. Within a few months market electricity prices had dropped well below what, in the midst of the crisis, had appeared to be justified. This multi-billion dollar mistake, borne by California taxpayers, was one reason Gray Davis lost his job as state governor to Arnold Schwarzenegger.

---

[196]Robert Crandall, "Air Pollution, Environmentalists, and the Coal Lobby," in Roger Noll and Bruce Owen, ed., *The Political Economy of Deregulation: Interest Groups in the Regulatory Process* (Washington: American Enterprise Institute, 1983), pp. 84–96.

[197]Milton Friedman, "Why Some Prices Should Rise," *Newsweek*, November 19, 1973, p. 130.

[198]Oliver Wendell Holmes, Jr., *Northern Securities Co. vs. United States*, 193 U.S. 197 at 400 (1904).

8. *Regulations often have unintended side effects.* New laws or regulations may change the incentives people face and encourage them to act in ways that the lawmakers had not foreseen.

   Recall the 1977 Clean Air Act amendments that placed strict emissions regulations on new power plants, while leaving existing facilities under the older standards (a practice known as "grandfathering"). Those rules increased the costs of new plants relative to existing ones, encouraging power companies to keep older plants in service longer than they otherwise would have been. Old plants are less efficient than new ones and the result was more fuel used and more pollution created.

   Fears of oil spills have led lawmakers to prohibit offshore drilling in many of America's coastal areas. As a result, the nation must import more oil than would otherwise be the case. However, imported oil is delivered via tanker. Tankers pose a greater oil spill danger than does offshore oil production. American coastlines are, therefore, actually less safe thanks to such legislative "protection."

9. *Regulators do not bear the costs of their regulations* and have little incentive to ensure that the benefits outweigh those costs.

10. *Public officials are self-interested, and their self-interest may not always be in the public interest.* As Nobel Prize-winning economist James Buchanan pointed out, "government policy emerges from a highly complex and intricate institutional structure peopled by ordinary men and women, very little different from the rest of us."[199] Buchanan and Gordon Tullock, the main developers of *Public Choice Theory*, argued that public officials have their own self-interests as much at heart as anyone else, and they may promote these interests at public expense.

    For instance, managers with the federal government are often paid in proportion to the number of people who report to them. Their incentive, therefore, is to increase the size of their departments. All too often, they act in accordance with this incentive regardless of the cost to taxpayers.

    More familiar are the politicians who purchase votes by using tax dollars to pay for projects of questionable value, or city officials who get kickbacks in return for construction contracts.

11. *Once in place, regulations are difficult to eliminate—the "tyranny of the status quo."*[200] Again, the example of oxygenates. Even though the problems of MTBE groundwater pollution and the increased gasoline evapora-

---

[199]James Buchanan, "From Private Preferences to Public Philosophy: The Development of Public Choice," in *The Economics of Politics* (London: Institute of Economic Affairs, 1978), p. 4.

[200]Milton Friedman, *The Tyranny of the Status Quo* (New York: Harcourt Brace Jovanovich, 1984).

tion caused by adding ethanol have been known for years, the regulations that require oxygenates have yet to be repealed. Oil companies are still required to put these harmful chemicals in their gasoline.[201]

No matter how harmful a regulation is, or how outdated it has become, there is usually someone who benefits by it. The beneficiaries of the regulation generally have a stronger interest in keeping the regulation in place than anyone else has in getting rid of it. As a result, they are willing to spend time and money lobbying the government to support their position. While the benefits of a regulation may be enjoyed by a relative few, the costs are often spread out among many. If the per person cost of a regulation is only a dollar or two, no one has a financial incentive to travel to Washington to lobby against it. Economists call this the *concentrated benefits and diffuse costs problem*.

Moreover, the benefits of any particular government action are usually quite visible while the costs are often hidden. For example, if the recycling industry receives a subsidy, the impact on that industry is very apparent in terms of new facilities and jobs. However, these gains may be more than offset by loss of facilities and jobs in other industries. Because of the taxes that must be raised in order to subsidize the recycling industry, consumers have fewer dollars with which to purchase goods and services from other companies. These losses, however, are diffuse and invisible.

Perhaps most importantly, people just do not like to admit when they have made a mistake, and politicians are no exception. If the "Smith Act" causes problems, Senator Smith is unlikely to apologize and propose that his Act be repealed. Instead, the senator will probably argue that his legislation was not properly funded or enforced. In the end, the law is more likely to be expanded than repealed.

12. *Industries exert enormous influence over the government agencies created to regulate them.* Reformers, believing this problem is due to an imbalance of power, often seek to remedy the situation by increasing the authority of the regulatory agency. Such measures will likely serve only to solidify the positions of those companies that already dominate the regulated business.

Industry sway over government agencies is a natural result of the incentives inherent in the regulatory process. As has already been pointed out, no one has more incentive to lobby regulatory agencies than do the companies they regulate. And regulators' self-interest gives them a powerful incentive to listen.

---

[201]The use of MTBE is being phased out over the next few years. However, other oxygenates, i.e., ethanol (currently the only practical alternative), will still be required.

There is also the "revolving door" phenomenon whereby person-
nel leave industry for jobs with government agencies and vice versa.
Some see this as proof of corruption, but there is a simpler and less
sinister explanation. When an agency is created to oversee a business, one of its first needs is employees with knowledge of that business. Where can it go for such people but to the industry itself?
Similarly, when government employees retire and wish to begin second careers, where can they go other than to the business about
which they have spent their professional lives learning?

13. *Laws and regulations stifle innovation.* Once a particular solution is writ-
    ten into law, there is little incentive for companies to try and develop
    a better one. Laws are notoriously difficult to change and are partic-
    ularly so when lobbyists' businesses depend upon the mandated so-
    lution. Even if the mandated solution was cutting-edge technology
    at the time the law was signed, technology becomes outdated very
    quickly in a free market system.

14. *National regulations can create nationwide problems.* In 1978, the Carter Ad-
    ministration, mistakenly convinced that the country was running out
    of oil and natural gas, passed the *Powerplant and Industrial Fuel Use Act.*
    Under the Act, existing power plants were prohibited from increasing
    their use of natural gas, and new plants were prohibited from using
    either natural gas or fuel oil. This restriction left coal as the only al-
    ternative despite the fact that coal emits more pollution and CO_2
    than does natural gas. President Reagan lifted the restrictions on ex-
    isting plants in 1981 and on new plants in 1987.

> **"**The art of economics consists in looking not merely at the immediate but
> at the longer effects of any act or policy; it consists in tracing the consequences
> of that policy not merely for one group but for all groups.**"**[202]
>
> **Henry Hazlitt, American journalist and economist**

. . . Or Setting Goals

Means-setting can pervert the goals. The objective ceases to be clean air,
clean water, or whatever other laudable end, and instead becomes adhering
to the *means* mandated by the regulation.

---

[202]Henry Hazlitt, *Economics in One Lesson,* p. 17.

Perhaps a better way to regulate is to simply define the goals and then get out of the way. That is:

1. Establish a goal (e.g., clean water).
2. Define a yardstick for determining whether the goal has been met (like specifying the maximum allowable levels of contaminants in wastewater that may be dumped into rivers, lakes, or oceans).
3. Establish the penalties for failing to meet the goal (e.g., monetary fines).
4. Let individuals and companies figure out how to meet the targets themselves.

People are amazingly creative. Given clear and reasonable goals, they will find ways to achieve them. And, with hundreds or thousands of people working towards a goal—trying different solutions, failing, then trying again, sharing information about what works and what does not—it is almost certain that their solutions will be far better than anything a regulator could devise.

The main problem with goal-setting is deciding what a reasonable target is. How clean is clean enough? Nowadays, chemical concentrations can be measured down to parts per trillion. Is water clean only if all measurable contaminants are removed? That can be done, but only at the cost of a lot of resources, energy, and pollution (from burning the fuel needed to power the contaminant removal process, disposing of the chemicals removed from the water, etc.).

Usually water is considered clean when contaminants are below levels that might cause harm to plant and animal life. This seems like a straightforward yardstick, but determining harm is anything but straightforward. For example, to determine whether a given chemical is carcinogenic (that is, if it will cause cancer), laboratory animals are typically fed the chemical at the highest non-lethal level (i.e., at amounts just below a dose that would kill them outright) for long periods. If the test animals develop cancer at a rate that is higher than normal, the chemical is considered to be a carcinogen.

There are many problems with identifying carcinogens in this way, however. Some chemicals are poisonous to some animals but not to others. In addition, chemicals that are toxic at high concentrations may actually be beneficial at lower levels (e.g., zinc, magnesium, and potassium). Reducing the concentration of such chemicals below their beneficial levels could actually be harmful to public health.

Also, there are *opportunity costs*. That is, when resources are used to make water absolutely pure, those resources are not available for other, perhaps more important, things. If billions of dollars are spent to reduce a pollutant to save an estimated ten lives per year, those dollars cannot be spent on highway improvements that could save a hundred lives a year. At what point do the costs exceed the benefits?

Clearly, it is important to balance the cost of cleaning the environment against the risk of leaving it less than perfectly clean.[203] Earlier in this chapter, we proposed a definition of pollution that included consideration of the volume of the pollutant and the ability of nature to deal with that volume. Perhaps a more practical definition of "clean" would allow emissions as long as they did not exceed a level that the local environment could handle.

CAP-AND-TRADE

A number of economists and environmentalists have championed a "market-based" alliance between government and industry to help clean the environment. Under this scheme, businesses would purchase the right to pollute. Certificates, known as pollution allowances, would grant the owner the right to emit a given quantity of a pollutant into the atmosphere each year. The total amount allowed by all the certificates issued would equal a level determined to be acceptable given local conditions.

These allowances could be bought and sold in the market like any other commodity. Companies could compare the price of buying allowances to the cost of reducing their emissions. Those companies able to reduce emissions for less than the market price of allowances would do so, and sell any unneeded allowances to others. Companies facing high emission control costs could purchase them instead.[204]

Environmental organizations could also buy certificates and remove them from circulation, thereby reducing the total amount of pollution allowable in their area.

Cap-and-trade combines the concept of government goal-setting with the market's ability to allocate resources to their best effect. Many economists believe that such a system would enable cities to control pollution far more efficiently than with traditional "command-and-control" regulations.

This scheme is not a cure-all. It offers an efficient mechanism to achieve an environmental goal, but the goal must be chosen with care. A job that is not worth doing is not worth doing efficiently. Moreover, once in place, cap-and-trade programs are difficult to eliminate. While such a program lasts, the pol-

---

[203]This issue can get very emotional. It is not unusual to hear people who advocate cleaning up the environment regardless of the cost ask, "How can we put a price on a human life?" The fact is we do just that every day of our lives. There is no doubt that a family would be safer if they drove, say, a Mercedes Benz instead of an economy car. However, they might reasonably choose a less expensive car so that they would have money to spend on food, clothing, shelter, and education for the children. By the same token, people who drive Mercedes might be safer if they drove Hummers, and Hummer drivers would be safer if they drove armored cars.

[204]John Swinton, "At What Cost Do We Reduce Pollution? Shadow Prices of SO_2 Emissions," *The Energy Journal*, Vol. 19, No. 1 (1998), pp. 66–67.

lution allowances have a monetary value that would disappear the moment the program ends. Participants that have a lot of money tied up in allowances would fight to keep the program going rather than lose their investment.

There could also be political problems if researchers were to determine that the environment could handle higher levels of a controlled substance than previously thought. Theoretically, such a finding should result in the issuance of more allowances. However, doing so would reduce the value of the certificates already in circulation, and would almost certainly create conflicts between people who already had allowances and those who needed them. Worse, one side or the other in such a debate might commission researchers to present misleading data in an effort to bolster their own position.

Some economists prefer emission taxes to cap-and-trade. Taxes are simpler to administer and easier to adjust or eliminate as conditions change or new information becomes available.[205]

HOW GREEN IS "GREEN?"[206]

Some power companies and independent marketers have begun offering *green energy*, or electric power generated from sources that are considered to be environmentally friendly. Consumers who choose to purchase such power pay a higher rate given that such energy is more expensive to generate. Some states, like California, help defray some of the costs to make green power more competitive with "nongreen" electricity.

The concept of green energy assumes that renewable technologies such as solar, wind, tidal, geothermal, and biomass have less environmental impact than do either hydrocarbon or nuclear power generation. Even though hydroelectric power produces no emissions, it is usually not considered green because it requires damming rivers and altering the local environment.

But are so-called green technologies really green? Windmills and solar panels provide only intermittent service, and conventional, nongreen, power sources must make up the difference when the wind is not blowing or the sun is not shining. Should solar and wind generation be considered less green because of this?

In addition, spinning wind turbine blades can kill birds (a Sierra Club official once described wind turbines as "Cuisinarts of the air"[207]). Should wind

[205]Bruce Stram, "A Carbon Tax Strategy for Global Climate Change," in Henry Lee, ed., *Shaping National Responses to Climate Change: A Post-Rio Guide* (Washington: Island Press, 1995), pp. 219–35.

[206]This section is adapted from Robert Bradley, Jr., "Green Pricing" in John Zumerchik, ed., 3 vols., *Macmillan Encyclopedia of Energy*, vol. 2, pp. 598–601.

[207]Paul Gipe, *Wind Energy Comes of Age* (New York: John Wiley & Sons, 1995), p. 450.

farms located in areas providing habitat for endangered species be rated lower than farms located in less sensitive regions?

Should geothermal energy be rated green, given that naturally occurring heat sources deplete with time, and some geothermal plants release toxic chemicals into the environment?

Should biomass be included as a green technology given that it produces air emissions and may encourage deforestation?

DIFFERENT ENVIRONMENTAL ISSUES

(Left) The body of a bird killed by a wind turbine at Spain's Tarifa wind farm. Tarifa successfully addressed its avian mortality problem, but other prominent wind sites like Altamont Pass wind farm, near San Francisco, continue to have problems with protected bird-species. *Source:* February 2, 1994 cover courtesy of Wind Power Monthly. (Right) An oil-soaked bird from a tanker spill. *Source:* Digitalvision.

On the other hand, should power from natural gas be added to the green list given that it is the cleanest of the fossil fuels and compares favorably with renewables on such measures as wildlife disturbance, noise, land use, and visual blight?

The definition of "green" will, and should, change with improving technology, regulatory reform, and new information about the environment. In the end, though, no power source is perfect; there will always be trade-offs.

ENERGY AND CLIMATE CHANGE

6

F UEL TO BURN?

As explained in Chapter 4, there are at least hundreds, and perhaps thousands, of years' worth of fossil fuels still available on Earth. However, the issue of man-made (anthropogenic) global warming has raised an important question: What will happen to the environment if that fuel is actually burned?

As the name implies, hydrocarbon molecules are made up of strings of hydrogen and carbon atoms.[208] When these molecules are *oxidized* (burned), heat is released along with water (H_2O), carbon monoxide (CO), and carbon dioxide (CO_2). Because air is about 79 percent nitrogen, oxides of nitrogen (NO_x) are also produced, and if the fuel contains sulfur, then sulfates (compounds of sulfur and oxygen) will be formed as well.

Carbon dioxide and nitrous oxide (N_2O) are *greenhouse gases* that, in high enough atmospheric concentrations, will warm the Earth's climate if no natural or human-driven processes offset the effect. Some scientists worry that such climatic changes might cause extreme heat and drought, more violent storms, higher ocean levels (putting coastlines and cities at risk), and increase the spread of tropical diseases.

Are these concerns justified, and if so, what can be done? The short answers are (respectively)"maybe" and "quite a bit."

THE GREENHOUSE EFFECT

Have you ever gotten into a car after it had been sitting in the sun and noticed how much hotter the air is inside the car than outside? This phenomenon is caused by the *greenhouse effect*. Sunlight passing through the car's windows is

---

[208]Coal is not a hydrocarbon. It is nearly all carbon and does not contain significant amounts of hydrogen (anthracite coal, for example, is 92 to 98 percent carbon).

absorbed by the interior, heating it and the air inside the car. Some of the heat passes back through the windows, but some is reflected off the windows back into the car. This trapped heat builds until the car's interior is warmer than the outside air.

The Earth's atmosphere acts like the car's windows, keeping heat from escaping back into space as infrared radiation. Greenhouse gases (such as carbon dioxide, methane, and water vapor) let incoming sunlight through, but block some of the infrared energy radiated upward by the sunlight-warmed Earth.[209] According to the EPA, "Without this natural greenhouse effect, temperatures would be much lower than they are now, and life as we know it would not be possible. Thanks to greenhouse gases, the Earth's average temperature is a hospitable 60°F,"[210] about 59°F warmer than it would be otherwise.[211]

But if the greenhouse effect becomes too strong, and if not enough radiated heat can escape from the atmosphere, then temperatures may rise too much.

Air pollution also has an impact on how much of the sun's energy penetrates the atmosphere and how much gets back out. Sulfates and particulates (e.g., smoke) may block the sun's incoming rays and therefore have a cooling effect.

When particulate emissions were much greater during the 1970s and 1980s, the possibility of global cooling was a concern.[212] Now that particulates are under better control, at least in western countries, global warming is the main worry.

GREENHOUSE GASES

As explained above, greenhouse gases are relatively transparent to visible light and relatively opaque to infrared radiation. They let sunlight enter the Earth's atmosphere, and, at the same time, keep radiated heat from escaping into space. The following sections provide brief descriptions of the most important of these gases.

Carbon Dioxide (CO_2)

By volume, carbon dioxide currently makes up 367 parts per million (0.0367 percent) of our atmosphere.[213] About 95 percent of this comes from natural

---

[209]Intergovernmental Panel on Climate Change, *Climate Change 2001: The Scientific Basis* (Cambridge, UK: Cambridge University Press, 2001), pp. 89–91.

[210]From the Environmental Protection Agency's web site: www.epa.gov/globalwarming/climate

[211]Patrick Michaels and Robert Balling, Jr., *The Satanic Gases: Clearing the Air About Global Warming* (Washington: Cato Institute, 2000), p. 25.

[212]James Fleming, *Historical Perspectives on Climate Change* (New York: Oxford University Press, 1998), pp. 131–36.

[213]IPCC, *Climate Change 2001: The Scientific Basis*, p. 185.

Digital Vision

sources (emissions from animal life, decaying plant matter, etc.) and the rest from human sources, mainly the burning of carbon-based fuels.[214]

While the human share of the total is relatively small, an estimated 3.5 percent to 5.4 percent, this additional contribution builds up over time because carbon dioxide is a very stable molecule and can last in the atmosphere for more than a hundred years. Since 1750, the atmospheric carbon dioxide concentration has increased by about 31 percent (from around 280 ppm) and is increasing at the rate of about 0.4 percent per year. It is estimated that carbon dioxide accounts for about 60 percent of the *anthropogenic* (or human caused) greenhouse change known as the *enhanced greenhouse effect*.[215]

If carbon fuels are of biologic origin, then sometime in the Earth's distant past there must have been far more CO_2 in the atmosphere than there is today. Over millions of years, much of it was removed by sea and land flora (plants). Most was returned to the air when the plant material decayed, but some of the carbon was locked up (or *sequestered*) in the form of wood, peat, coal, petroleum, and natural gas. Now that we are burning these fuels, the carbon is being released into the atmosphere once again.

[214]Ibid., p. 121.

[215]Ibid., p. 7, and Tom Wigley, "The Science of Climate Change," in Eileen Claussen, ed., *Climate Change: Science, Strategies, & Solutions* (Boston: Brill, 2001), p. 70.

Carbon dioxide is less soluble in warmer water than in cold, and as ocean surface layers warm, CO_2 could be driven out of solution and into the atmosphere, thus exacerbating the problem.[216]

There are benefits to increased CO_2 concentrations as well as potential problems. Plants need carbon dioxide; in fact, the optimal concentration for most plants is estimated to be between 800 and 1,200 ppm.[217] Some plants do best at even higher concentrations; the optimal range for rice is 1,500 to 2,000 ppm.[218] As the atmosphere becomes richer in CO_2, crops and other plants will grow more quickly and profusely. A doubling of carbon dioxide concentrations can be expected to increase global crop yields by 30 percent or more.[219]

Higher levels of CO_2 increase the efficiency of photosynthesis, and raise plants' water-use efficiency by closing the pores (stomates) through which they lose moisture. Carbon dioxide's effect is twice that for plants that receive inadequate water than for well-watered plants. In addition, higher CO_2 levels cause plants to increase their fine root mass, which improves their ability to take in water from the soil.[220] Higher water efficiency should allow plants to better cope with hotter climates.

Water Vapor (H_2O)

The most common greenhouse gas is *water vapor*, which accounts for about 94 percent of the natural greenhouse effect.[221] Its atmospheric concentration is ten times that of CO_2.

Water vapor's impact on the climate is complex and not well understood. It can both warm and cool the atmosphere. When water evaporates, it cools the surface from which it evaporates. In addition, heavy clouds block sunlight and reflect it back into space.

On the other hand, thin cirrus clouds may tend to let solar energy in while keeping radiated energy from escaping into space. Also, moist air retains more heat than does dry air, so a humid atmosphere should be warmer than a dry one. On balance, it is believed that water vapor has a net warming effect.

---

[216]IPCC, *Climate Change 2001: The Scientific Basis*, p. 200.

[217]Sylvan Wittwer, *Food, Climate, and Carbon Dioxide: The Global Environment and World Food Production* (Boca Raton, FL: CRC Press, 1995), p. 101.

[218]Ibid., p. 113.

[219]Ibid., 86.

[220]Patrick Michaels and Robert Balling, Jr., *The Satanic Gases*, p. 184.

[221]Ibid., p. 25.

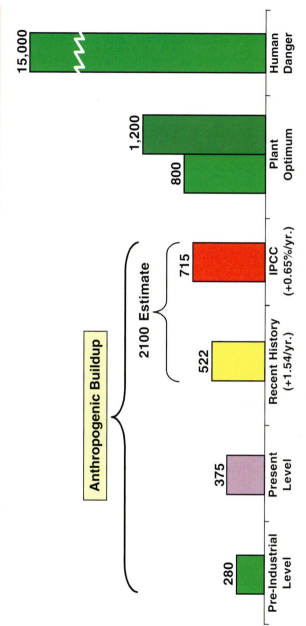

Atmospheric CO₂ & Social Welfare (parts per million by volume)

The positive effects of carbon fertilization on crop yields lead some economists to suggest that the benefits of higher atmospheric CO_2 concentrations may outweigh the costs for decades to come. *Source: Climate Change 1995—The Science of Climate Change,* pp. 21–26 by John Houghton, ed.; *Food, Climate, and Carbon Dioxide* pp. 89–91 by Sylvan Wittwer; *In Defense of Carbon Dioxide,* p. 1 by New Hope Environmental Services.

Pre-Industrial Level — 280

Present Level — 375

Recent History (+1.54/yr.) — 522

IPCC (+0.65%/yr.) — 715

Plant Optimum — 800, 1,200

Human Danger — 15,000

Anthropogenic Buildup

2100 Estimate

The main concern about increased concentrations of atmospheric water vapor is the possibility of a strong *positive feedback effect*.[222] As the climate warms, more water will evaporate, increasing the amount of water in the air. The increased concentration will, in turn, further warm the climate leading to a still higher level of water vapor in the atmosphere. This iterative cycle, it is feared, could spiral out of control, resulting in damaging or even catastrophic temperature increases.

However, if the Earth's climate were that sensitive, it would probably have spun out of control long before now given that there have been periods in the distant past when temperatures and CO_2 were higher than they are today. This leads some scientists to suspect that there may be natural mechanisms working to keep the climate in balance.

Meteorologist Richard Lindzen has proposed one such mechanism, which he calls the Iris Effect.[223] Lindzen and his colleagues suggest that upper-level cirrus clouds, which tend to trap heat radiated from the Earth's surface, may open "as an iris (by analogy with the eye's iris)" in response to higher earth surface temperatures.[224] Lindzen's Iris Effect is only a hypothesis, but important scientific work is beginning to suggest that the water vapor feedback is not as strongly positive as indicated by some computer climate-models.

Methane (CH_4)

While methane is 25 times more powerful a warming agent than carbon dioxide, it has a much shorter life span and its atmospheric concentration is only about 17 ppm. Concentrations have more than doubled since 1850, though for reasons that are still unclear, they have leveled off since the 1980s.[225] Human activity accounts for about 60 percent of methane emissions, while the rest comes from natural sources such as wetlands. Human sources include leakage from pipelines, evaporation from petroleum recovery and refining operations,

---

[222]The term "feedback effect" was borrowed from the name given to the loud screech that is produced when a microphone is placed near a speaker. Background noise, picked up by the microphone, is amplified and then fed to the speaker. The microphone picks up this amplified sound, sends it back to the amplifier, and so on until the noise becomes painfully loud. The word "positive" in the phrase, "positive feedback effect," is not meant to imply "good." Rather it means that the effect builds on itself, growing larger and larger. A *negative feedback effect* would be one that tends to dampen itself out.

[223]Lindzen is currently the Alfred Sloan Professor of Meteorology in the Department of Earth, Atmospheric, and Planetary Sciences at MIT. He is widely respected and highly credentialed (Lindzen was one of the youngest members ever elected to the National Academy of Sciences), but, while other scientists researching global warming consider him brilliant, they also see him as something of a maverick.

[224]Richard Lindzen, Ming-Dah Chou, and Arthur Hou, "Does the Earth Have an Adaptive Infrared Iris?," *Bulletin of the American Meteorological Society*, September 2001, p. 417.

[225]E. J. Dlugokencky, et al., "Atmospheric Methane Levels off: Temporary Pause or a New Steadystate?" *Geophysical Research Letters*, October 8, 2003, doi:10.1029/2003GL018126, 2003.

rice fields, coal mines, sanitary landfills, and wastes from domestic animals. About 20 percent of the total human greenhouse impact is due to methane.

Nitrous Oxide (N_2O)

Nitrous oxide's warming potential is some 300 times that of CO_2. It has an atmospheric concentration of about 0.32 ppm, up from 0.28 ppm in 1850. In the United States, 70 percent of man-made nitrous oxide emissions come from the use of nitrogen-containing agricultural fertilizers and automobile exhaust. Globally, fertilizers alone account for 70 percent of all emissions.

Catalytic converters, whose use on car exhaust systems was federally mandated in 1970 by the Clean Air Act, increase N_2O emissions, though to what extent is under debate. The EPA has calculated that production of nitrous oxide from vehicles rose by nearly 50 percent between 1990 and 1996 as older cars without converters were replaced with newer, converter-equipped models. Critics argue that the EPA's numbers are greatly exaggerated. In addition, they point out that converters reduce emissions of another greenhouse gas, ozone, as well as carbon monoxide and NO_x (which leads to smog).

CFCs

Chlorofluorocarbons, or CFCs, are powerful global warming gases that do not exist in nature but were invented by scientists at an American chemical company in the 1930s. They are used as propellants in aerosol sprays and as refrigerants. Freon, the most well-known CFC, was widely used in refrigerators and in home and auto air conditioning systems until it was banned in 1995.[226]

The fact that CFCs are chemically inert (that is, they do not react with other chemicals) makes them very useful in a wide variety of applications, but it also means that they last for a very long time in the atmosphere (perfluoromethane, for example, can persist for 50,000 years).

These gases affect the climate in different ways depending upon their location in the atmosphere. At lower altitudes, they trap heat like other greenhouse gases and have a much stronger warming effect than CO_2. In fact, some can trap as much as 10,000 times more heat per molecule than carbon dioxide. While CO_2 is measured in atmospheric concentrations of parts per million, CFCs are measured in parts per trillion. Despite their low concentrations, it is believed that these gases account for about 15 percent of the human greenhouse change.

[226]Freon's use was originally scheduled to be phased out by the year 2000 in accordance with the 1987 Montreal Protocol, but the timetable was advanced in response to pressure from environmentalists. Developing countries and Eastern-bloc nations did not sign the protocol, and still use the chemical. In addition, Freon is being smuggled into the United States because it generally costs less, works better, and is less toxic and corrosive than its replacements.

In the upper atmosphere, or stratosphere, chloroflourocarbons are broken down by sunlight. The chlorine that is released by this decomposition acts as a catalyst to break naturally occurring ozone (O_3) molecules into oxygen (O_2) molecules. Ozone helps block the sun's ultra-violet radiation, which can cause skin cancer after long-term exposure.

Worldwide CFC emissions have been steadily dropping, and it is expected that ozone depletion (the ozone hole), which reached its peak in the last decade, will drop to zero later this century.[227]

IS THE CLIMATE WARMING?

We know that concentrations of carbon dioxide, methane, and nitrous oxide are increasing due to human activity, but is the climate getting warmer as a consequence? The evidence, while still not conclusive, suggests that it is. A number scientists point to ground measurements taken over a number of decades that indicate a noticeable temperature rise.

On the other hand, skeptics point out that the data are skewed toward urban areas where most measurements are taken. The problem, they argue, is that cities tend to be warmer than the surrounding countryside because of heat absorption by streets, parking lots, and dark roofs. While measurements are adjusted to compensate for this *urban heat island effect*, the critics claim that the adjustments are insufficient.

In addition, as Russia's economy worsened, the country stopped taking ground-based measurements in many areas. As a result, data from cooler regions of the globe have been significantly reduced.

Another criticism is that methodical, direct atmospheric temperature monitoring has only begun in recent decades. Estimates of past temperatures are based on observations of coral reefs and tree rings. Such indirect measurements are subject to uncertainty, and any trend analysis based on such data is open to question.

Furthermore, temperature readings from satellites and weather balloons are not detecting the *greenhouse signal* in the area where it should be strongest—the lower troposphere.[228] It is possible, though, that ozone depletion or other factors could be producing a downward bias in these readings.[229]

---

[227]U.S. Environmental Protection Agency, *Latest Findings on National Air Quality: 2001 Status and Trends* (Washington: EPA, 2002), p. 240.

[228]The troposphere is that part of the atmosphere that extends from the earth's surface to an altitude ranging from about five miles over the polar regions to eight miles over the equator.

[229]Robert Bradley, Jr., *Julian Simon and the Triumph of Energy Sustainability*, p. 98.

GLOBAL TEMPERATURE DISCREPANCIES

This graph compares the global average temperature increase (1979–2003 in F°/decade) over the last quarter century as measured by three sets of surface readings vs. the average temperature measured by satellites and balloons. The discrepancy between the two sets of measurements is currently an active area of research. The last set of bars indicates the temperature increase predicted by climate models. *Source:* Goddard Institute for Space Studies (NASA), National Climate Data Center, and Intergovernmental Panel on Climate Change.

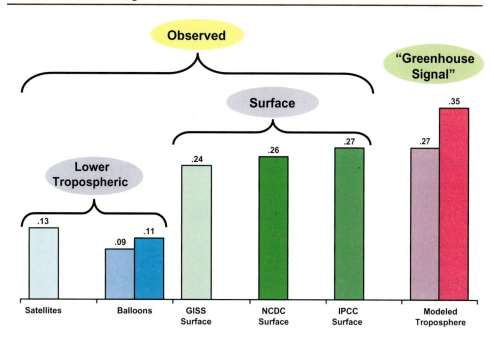

At least some of the last century's warming is believed to be the result of increased solar activity (the sun's intensity rises and falls over time with 11-, 22-, and 88-year cycles). One study suggests that "the increase in direct solar irradiation over the last 30 years is responsible for about 40 percent of the observed global warming."[230] Changing ocean currents may have also produced a warm phase coming out of the little ice age in the mid-nineteenth century.

Nonetheless, according to the Intergovernmental Panel on Climate Change (IPCC), temperature readings at ground-based measuring stations reveal an average warming trend of about 1.1°F (0.6°C) since 1850 after adjusting for the urban heat island effect. About half of this warming has occurred since 1970, which, to many scientists, is proof of an emerging greenhouse signal.

[230]For a summary of solar/climate studies, see Bjørn Lomborg, *The Skeptical Environmentalist*, pp. 276–78.

In 1988, the *Intergovernmental Panel on Climate Change* was created under joint sponsorship of the United Nations and the World Meteorological Organization (WMO). While the IPCC is not itself a scientific research organization, its three working groups—WG I Science, WG II Impacts and Adaptation, and WG III Mitigation—each issue a report every five years on the findings of the latest climate-change research. The First Assessment Report was completed in 1990, the second in 1995, and the third in 2001. The Fourth Assessment Report is scheduled for completion in 2005.

Along with the book-length reports produced by each of the three working groups, the IPCC issues a 20-page *Summary for Policymakers*. This summary is politically influential and is usually the only part of the report that gets read.

Though it is generally acknowledged that the IPCC reports present the best available science, critics argue that the executive summaries tend to be much more alarmist than the scientific portions of the reports warrant.[231]

COVERS OF **IPCC** SCIENCE REPORTS

The IPCC's second and third scientific assessments, released in 1995 and 2001, respectively, have provoked the greatest international debate on energy usage and policy in history. *Source:* IPCC.

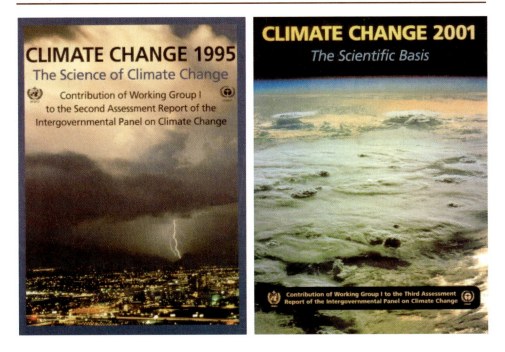

---

[231]According to the National Research Council's report, *Climate Change Science, An Analysis of Some Key Questions* (Washington: National Research Council, 2001, p. 4), "The *Summary for Policymakers* reflects less emphasis on communicating the basis for uncertainty and a stronger emphasis on areas of major concern associated with human-induced climate change."

HAVE THERE BEEN OTHER CHANGES?

Along with higher temperatures, many scientists expect that anthropogenic global warming will mean a more active *water cycle* (increased evaporation and rainfall, for example) and higher sea levels (due to thermal expansion of the oceans' water and to melting ice sheets). Some scientists argue that more extreme weather events such as tornadoes, hurricanes, dust storms, and droughts could also occur.

At this time, there is not enough information to identify trends in these areas with much assurance. Data quality is poor, incomplete, and of limited duration. For example, measurements of ice sheet changes are contradictory, and we do not know whether they are growing or shrinking. While the melting of floating ice (such as that at the Arctic) will not cause sea levels to rise, the land-based ice masses in Greenland and Antarctica are of concern. However, the IPCC predicts that in the 21st century, increased water runoff in Greenland (from warming) will be offset by an ice buildup in Antarctica (from more precipitation).[232]

The best available research to-date indicates the following:[233]

- There is no firm evidence of a global increase in extreme weather events during the 20th century.
- There has been a slight increase in rainfall of about 1 percent in the Northern Hemisphere.
- Sea levels have risen over the past hundred years; estimates range from about 4 to 10 inches (10 cm to 25 cm).
- The temperature of the top 10,000 feet of the ocean has risen approximately 0.11°F (0.06°C) between 1955 and 1996.[234]
- Between 1955 and 1996 there was also an increase in the average temperature of the top 1,000 feet of the ocean. Strangely, however, this rise appears to have all occurred during the years 1976 and 1977. Climatologists call this phenomenon "the great Pacific climate shift," but have not been able to explain it.[235]

[232]IPCC, *Climate Change* 2001: *The Scientific Basis*, pp. 650–52, 664–65, and 668–70.

[233]Unless otherwise noted, this information is taken from IPCC, *Climate Change* 2001: *The Scientific Basis*, pp. 4–5, 11, 15–16, 33, 73, 104, 641, and 699.

[234]Sidney Levitus et al., "Warming of the World Ocean," *Science*, March 2000, p. 2227.

[235]The climate shift that some scientists believe occurred in 1976–1977 may not have been a natural phenomenon at all, but may have only been a shift in the data caused by the closing of Soviet monitoring stations.

ARE PEOPLE CAUSING THESE CHANGES?

The human contribution to these trends is uncertain because there is so much natural climate variation. Not only are there solar activity cycles, but there is also a 100,000-year ice age cycle. Currently, we are enjoying one of the cycle's 10,000- to 30,000-year warming periods. As James Hansen pointed out, "Climate is always changing. Climate would fluctuate without any . . . [man-made] climate forcing. The chaotic aspect of climate is an innate characteristic."[236] In Richard Lindzen's opinion, "[W]e are not in a position to confidently attribute past climate change to carbon dioxide or to forecast what the climate will be in the future. . . . One reason for this uncertainty is that . . . the climate is always changing; change is the norm. Two centuries ago, much of the Northern Hemisphere was emerging from a little ice age. A millennium ago, during the

COOLING AND WARMING BOOK COVERS

Between the mid-1940s and mid-1970s, global surface temperature readings, indicating that the climate was cooling, sparked a fear that Earth was headed toward another little ice age. When the trend reversed soon thereafter, the concern became global warming. *Source:* (left) Illustration by Jean Spitzner (right) Cover design by Lawrence Ratzkin.

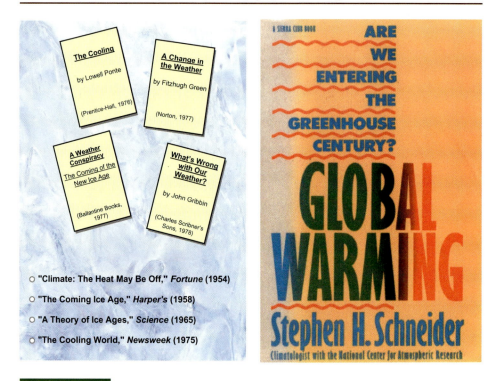

○ **"Climate: The Heat May Be Off,"** *Fortune* (1954)

○ **"The Coming Ice Age,"** *Harper's* (1958)

○ **"A Theory of Ice Ages,"** *Science* (1965)

○ **"The Cooling World,"** *Newsweek* (1975)

---

[236]James Hansen, "How Sensitive is the World's Climate?," *National Geographic Research & Exploration*, Vol. 9, No. 2 (1993), p. 143.

Middle Ages, the same region was in a warm period. Thirty years ago, we were concerned with global cooling."[237]

Despite the uncertainty, the IPCC's 1995 report concluded, "The balance of evidence suggests a discernible human influence on global climate." Its conclusion was based on mathematical analysis "reality checked" with available data. The IPCC hedged by stating: "Although these global mean results suggest that there is some *anthropogenic* (i.e., human-made) component in the observed temperature record, they cannot be considered as compelling evidence of a clear cause-and-effect link between anthropogenic forcing and changes in the Earth's surface temperature."[238]

The 2001 report concluded that "there is new and stronger evidence that most of the warming observed over the last 50 years is attributable to human activities."[239]

Despite the IPCC's increasing confidence, Richard Kerr of *Science* pointed out that the range of uncertainty as reflected in the data and projections presented in the 2001 report actually *increased* in comparison with that presented in the earlier report.[240] The more we learn, the more we learn that there is more to learn.

COMPUTER MODELING

The temperature increase thus far detected is significantly less than the computer model predictions that have been the source of much of the concern over global warming. While these models are very sophisticated and run on extremely powerful machines, they have not yet been able to accurately mirror the immensely complex greenhouse that is Earth. Clearly, there are still many unknowns that must be resolved and much more data that must be collected before the models can be trusted.

Some of the unknowns are:

- The effect of the oceans, which may be acting as *heat sinks* absorbing some of the heat that would otherwise be raising atmospheric temperatures. If so, then the oceans' ability to absorb heat may drop as it gets warmer (heat transfer rate is proportional to temperature difference).
- The impact of plant life, both on land and in the sea. Higher carbon dioxide levels will lead to increased plant life and greater crop yields.

[237]Richard Lindzen, "Scientists' Report Doesn't Support the Kyoto Treaty," *Wall Street Journal*, June 11, 2001, p. A22.

[238]IPCC, *Climate Change 1995: The Science of Climate Change* (Cambridge, UK: Cambridge University Press, 1995), p. 4.

[239]*Climate Change 2001: The Scientific Basis*, p. ix.

[240]Richard Kerr, "Rising Temperature, Rising Uncertainty," *Science*, April 13, 2001, pp. 192–94.

More plants will mean more carbon removed from the atmosphere—though not enough to completely offset fossil fuel emissions.

- The water vapor feedback effect. Is it positive or negative, and what is its magnitude?
- The effect of clouds.
- The contribution of natural climate variability, including solar activity.
- The net impact of aerosols (i.e., suspended particles). Some of these particulates reflect light and tend to cool the atmosphere. Others absorb light and can have a warming effect. Most aerosols (about 90 percent) such as dust from soil, volcanic dust, and sea salt are of natural origin. Aerosols of human origin include soot and sulfates from burning carbon-based fuels.

Climate scientists are working to resolve the issues and collect needed data. However, as the technical summary of the IPCC's 2001 report states, "The climate system is a coupled nonlinear chaotic system, and therefore, the long-term prediction of future exact climate states is not possible."[241] (For a plain-English explanation of why this is so, refer to *Appendix C—The Butterfly Effect*.)

Critics of climate modeling point out that meteorologists cannot even accurately predict the weather more than three or four days in advance. How then, they ask, can modelers hope to predict the climate 50 or 100 years from now? While there is some merit to this argument, the fact is that predicting climate change is not the same as predicting the weather. It is both simpler and more complex.

It is simpler in that climate scientists do not have to determine specific future weather conditions (for example, if there will be a tornado in Wichita Falls, Texas, on June 6, 2050). Instead, they are trying to identify trends in the world's climate—a global "average" of local weather conditions. On the other hand (as explained in *Appendix C*), in any iterative, nonlinear system, small and immeasurable causes may have huge effects over long periods.

There are many pitfalls, and many opportunities for human error. As pointed out by climatologist Gerald North, climate modelers can be the victims of a "group think" bias. That is, scientists have a tendency to "calibrate" or adjust their models to agree with other models. Few researchers want to publish predictions that are either significantly higher or lower than the norm. Nor do agencies that fund this research want to pay for results that are outside the mainstream.[242]

The problems inherent with attempting to model Earth's atmospheric mechanisms make quantitative predictions suspect. Most likely we will only know what the climate will be like 50 years from now in 50 years.

---

[241]IPCC, *Climate Change 2001: The Scientific Basis*, p. 78.

[242]Gerald North, book review, "L. Danny Harvey, *Global Warming: The Hard Science*," *Climate Change*, June 2001, pp. 293–97.

That said, the fact remains that all the models and the available data do point to a warming trend. The following diagram shows the projected increase in average climatic temperature if CO_2 concentrations in the atmosphere double, triple, or quadruple over the coming centuries. Two scenarios, one assuming neutral climate feedback effects and the other assuming strong positive feedback (which the IPCC models predict), are compared to an estimated "problematic" warming level.[243]

The importance of feedback effects on global warming is shown in this graph. In the neutral feedback case, the amount of projected warming is tolerable. However, given strong feedback effects, warming, as estimated by the IPCC, could reach a level of concern before the end of this century.

ATMOSPHERIC GHG BUILDUP & WARMING

Scientists agree that temperatures will rise with increasing atmospheric concentrations of CO_2. The questions are, "how much?" and "how bad?" The range between the "best" and "worst" case scenarios (in comparison with the estimated level at which the net effect of warming will be negative) illustrate the uncertainty that currently surrounds the issue. *Source:* See footnote 243.

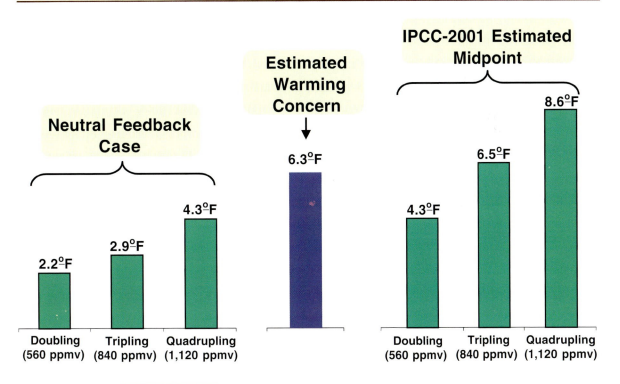

[243]IPCC, *Climate Change 2001: The Scientific Basis*, p. 577; Steve Schneider, "What Is 'Dangerous' Climate Change?" *Nature*, May 3, 2001, pp. 17–19.

ANTHROPOGENIC SURFACE
WARMING
Data collected from 1951 through 1997 shows anthropogenic warming is greatest in the colder regions of the globe.[244] Warming is also skewed toward the coldest times of the year in these regions. *Source:* James Hansen et al., "A Closer Look at United States and Global Surface Temperature Change," *Journal of Geophysical Research,* volume 106, pp. 23, 947–53, 963.

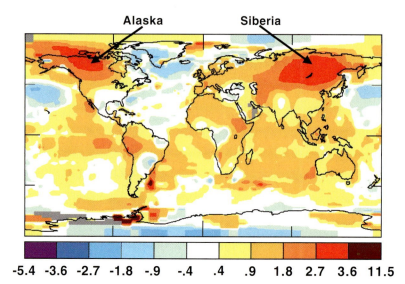

Alaska   Siberia

-5.4 -3.6 -2.7 -1.8 -.9 -.4 .4 .9 1.8 2.7 3.6 11.5

WARMING DISTRIBUTION

Actual data and theoretical modeling indicate that this warming trend disproportionately affects lower temperatures and frigid regions during the coldest times of the year. Over the past 40 years, nights have warmed more than days, with minimum temperatures typically increasing twice as much as maximum temperatures. The above chart, created by NASA meteorologist James Hansen, shows that the areas of greatest warming are concentrated in Siberia and Alaska.

The fact that global warming is greatest in colder regions mitigates its effects somewhat. In fact, colder areas will likely benefit from a warmer climate. Economic impact studies suggest that North America, Europe, and the former Eastern-bloc countries may gain from moderate global warming—primarily due to longer growing seasons.[245]

However, while warming may be concentrated in colder regions, it is not confined there. Areas that already have very warm climates such as Africa, India, Mexico, and Central America will likely suffer disproportionately from the trend. Similarly, while rising sea levels may be only a matter of inconvenience to some countries, they could pose a serious problem to island nations.

[244]Robert Balling, et al., "Analysis of Winter and Summer Warming Rates in Gridded Temperature Time Series," *Climate Research,* February 27, 1998, p. 178.

[245]Robert Mendelsohn, *The Greening of Global Warming* (Washington: American Enterprise Institute, 1999), p. 25. Mendelsohn's conclusions made in 1999 have not significantly changed as of 2004 (communication from Mendelsohn to authors, February 21, 2004).

ESTIMATED WARMING EFFECTS: YEAR 2100

Climate economist Robert Mendelsohn has estimated the economic impact of IPCC-predicted warming, precipitation, and sea level rise over the coming century for different regions of the world. The regions shown in green are expected to gain from climate change, while those in red will be hurt. These differences stem from each region's current climate and level of wealth. Areas that are currently the warmest will be hurt the most, and the poorest regions will be least able to adapt. Mendelsohn expects that globally, the costs and benefits will balance out. *Source:* Robert Mendelsohn. *The Greening of Global Warming,* p. 18. Reprinted by permission of Robert Mendelsohn.

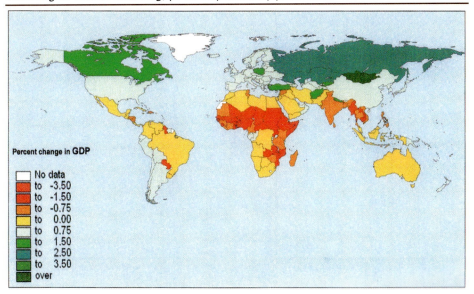

Percent change in GDP

	No data
	to -3.50
	to -1.50
	to -0.75
	to 0.00
	to 0.75
	to 1.50
	to 2.50
	to 3.50
	over

Whatever the impact of global warming, wealthy nations will be better able to adapt than will poorer ones.

SOLUTIONS

A number of possible methods of dealing with climate change have been proposed. These fall into three main categories:

1. Prevention:
   - Reduce carbon dioxide emissions by producing and using energy more efficiently.[246]
   - Develop carbon-free energy sources such as nuclear power and renewables.

---

[246]Increased efficiency can be a "double-edged sword." As power generation efficiency increases, energy prices drop and consumption rises.

- Use zero-emission coal technology to extract hydrogen from water reacting with coal.[247] Capture and sequester any CO_2 produced in the process.[248]
- Remove carbon dioxide from plant flue gas emissions to make fertilizer, or to produce chemical products. The extracted CO_2 could also be injected into petroleum reservoirs to help increase oil production.
- Work to prevent soil erosion in order to keep carbon material sequestered.
- Switch to no-till agriculture, which can reduce soil erosion by as much as 98 percent.
- Change rice cultivation methods to reduce methane emissions.
- Reduce methane emissions by working to eliminate pipeline and drilling leaks.
- Use existing technology to prevent methane leaks from landfills.

2. Correction:
- Expand *carbon sinks* (e.g., forests).
- Remove CO_2 directly from the atmosphere.
- Increase the amount of sulfates in the atmosphere (perhaps by adding sulfur to jet fuel in commercial airliners) to reflect sunlight.

3. Adaptation:
- Increase world trade to improve the productivity and wealth of poorer nations so that they have the resources to deal with problems that may arise from a changing climate.
- Build dikes and seawalls to block rising seawater.
- Migrate from warmer to cooler climates.

People's self-interest will drive them toward solutions. For example, power companies are cutting costs by investing in new technologies and using natural gas instead of coal. Combined-cycle natural gas plants are more efficient than traditional oil- and coal-fired plants, and natural gas burns cleaner and produces less carbon dioxide (30 percent less than oil and 40 percent less than coal per unit of energy generated).

Between just the early eighties and nineties, the efficiency of new power plants increased by 50 percent. Government regulations could encourage, or at least not discourage, the construction of new plants. Consider, for instance, that 45 percent of the cost of New Mexico's San Juan coal-fired power plant

[247]Han-Joachim Ziock, Klaus Lackner, and Douglas Harrison, "Zero Emission Coal," *Energy* 2000, pp. 268–274.

[248]Martin Hoffert, et al., "Advanced Technology Paths to Global Climate Stability," p. 983.

was spent in complying with environmental regulation.[249] Some economists have pointed out that less stringent regulations on the construction of new plants could actually result in a cleaner environment (and fewer CO_2 emissions) by making it more likely that older, dirtier, and less efficient plants get upgraded or replaced by newer (though less than perfectly clean) plants.

Refineries and other plants that are not normally in the power business can build *cogeneration* facilities that convert waste heat into electrical power. They can use this power themselves or sell it on the market.

Offices can cut power costs by using more efficient lighting (e.g., fluorescent rather than incandescent), and painting their buildings' roofs white to reflect heat (studies have shown that white roofs are significantly cooler than black ones, and can lower air conditioning costs by up to 40 percent).[250]

Some scientists and policy makers are urging the government to promote the use of nuclear power as a replacement for fossil fuels. Nuclear plants produce no carbon dioxide. In fact, it is estimated that the world's nuclear power plants save 550 million metric tons of carbon in the form of CO_2 from being released into the atmosphere each year.[251] As discussed in Chapter 2, however, atomic plants must confront serious waste disposal issues (both technical and political). Electric power from fusion may be a long-term solution, but it is probably decades away.[252]

Electricity can also be produced by such zero-or-low emission technologies as solar cells, wind turbines, tidal power, and geothermal generators, but each of these sources suffers from the limitations discussed in the second chapter.

Interest in using hydrogen as an automobile fuel has been growing. Hydrogen is very efficient and clean; burning it produces only water and some nitrogen oxides. However, hydrogen is very reactive and does not exist in a free state on Earth. Hydrogen is, therefore, not a primary energy source. It can be generated by water hydrolysis, a process that consumes a lot of electrical power. Hydrogen can also be extracted from hydrocarbon fuels. However, these processes generate less energy (in the form of hydrogen) and more CO_2 than would be produced by burning the hydrocarbons directly.[253] On the other hand, CO_2 produced at a central plant would be far more easily contained and sequestered than CO_2 emitted by thousands of individual gasoline-burning automobiles.

[249]Christopher Flavin and Nicholas Lenssen, *Power Surge*, p. 67.

[250]John Emshwiller, "California's Shortages Rekindle its Efforts to Conserve Electricity," *The Wall Street Journal*, February 20, 2001, p. A1.

[251]*New Energy Technologies: A Policy Framework for Micronuclear Technology*, Baker Institute for Public Policy, Rice University, Houston, Texas, September 2001, p. 3.

[252]D. W. Ignat, "Nuclear Fusion" in John Zumerchik, ed., *Macmillan Encyclopedia of Energy*, vol. 2, p. 878.

[253]Martin Hoffert, et al., "Advanced Technology Paths to Global Climate Stability," p. 983.

Microbiologists are currently working with genetically engineered bacteria that can efficiently convert waste biomass into ethanol for fuel. From the standpoint of carbon dioxide emissions, there might appear to be little difference between burning alcohol and burning gasoline. However, burning gasoline releases new carbon dioxide into the atmosphere, while burning ethanol only puts back what was there before (that is, before it was removed from the atmosphere by the plants that produced the biomass products).

Telecommuting, or logging into work via computer rather than driving in by car, could reduce CO_2 emissions along with other pollutants. Computers can help in countless other ways as well, not least by enabling people to easily share information and ideas over the Internet.

An obvious way to help offset CO_2 emissions is to plant trees. One much debated paper has suggested that North America may already have enough trees to absorb all the carbon dioxide that the continent emits.[254] However, unless trees are harvested and their wood turned into relatively permanent objects (houses, furniture, etc.), the amount of carbon that our forests can absorb will reach equilibrium at some point. The CO_2 absorbed will be balanced by the amount given off by burning or decaying trees.

The 1990s policy of favoring wilderness areas over managed forests reduced the effectiveness of national parks in serving as carbon sinks. Managed forests can support many more trees per acre than can wildernesses, and trees in wilderness areas eventually decay or burn so that their carbon is ultimately returned to the atmosphere. The choice between the carbon sink capacity of managed forests and scenic, untouched wilderness is another difficult environmental trade-off.

The Kyoto Protocol is an attempt to limit CO_2 emissions through an international cap-and-trade scheme. Though finalized in 1997, not enough countries ratified the protocol to bring it into force. The agreement, which does not have the support of the United States, would obligate the 38 developed countries to reduce their aggregate GHG emissions by 5 percent from 1990 levels by 2008–2012. Under the agreement, the developing nations are not covered by mandatory, or even voluntary, measures.

The protocol got off to a slow start because it was perceived by many to be all pain and no gain. It has been calculated that if all the nations met their obligations under the protocol by 2050, the reduction in the amount of global warming would be 0.13°F (0.07°C), an amount impossible to detect given the

[254]S. Fan et al., "A Large Terrestrial Carbon Sink in North America Implied by Atmospheric and Oceanic Carbon Dioxide Data and Models," *Science*, October 16, 1998, p. 442.

natural temperature swings from year-to-year.[255] The cost of this insignificant gain would amount to trillions of dollars in resources.[256]

Even many of its supporters in the scientific community admit that the protocol would do little to halt global warming, but they still back the treaty as an important first step in the right direction. Critics point out, however, that adhering to the treaty would have a significant impact on many nations' economies. If the impact were severe enough, a second step could be politically impossible.

Worse, though, is the incalculable loss of those trillions of dollars worth of resources spent for so little benefit. If those resources were used instead to create real wealth, countries that are now poor would be better able to adapt to a changing climate. They could also be invested to develop technology capable of solving the problem far more cheaply and effectively.

Another danger of an international cap-and-trade plan like Kyoto is that it provides an incentive for dictators to retard the economic growth of their countries so that they can sell unused carbon emission credits to other nations. Money flowing from democracies to tyrannies in this manner would only serve to prop up corrupt and despotic regimes.

Often a country's motives in signing a treaty like Kyoto are not obvious. When the United States refused to ratify the Kyoto accord, Europeans expressed great moral outrage (despite the fact that only one European country, Romania, had ratified it at that time). However, Margaret Wallström, the environment commissioner of the European Union, tacitly admitted that one reason for the anger was that Europeans had hoped that the treaty would weaken American manufacturing.

"This is not a simple environmental issue," she stated. "This is about international relations, this is about [the] economy, about trying to create a level playing field for big businesses throughout the world. You have to understand what is at stake and why it is serious."[257]

Energy costs are much greater in Europe than in the United States largely because of substantially higher levels of taxation and regulation. As a result, American companies have a significant advantage over their European competitors. Kyoto would have disproportionately driven up energy costs in the United States and helped to reduce this advantage.[258]

When participants in such negotiations have "an ideological axe to grind or a financial stake in the outcome, or both,"[259] scientific objectivity is quickly lost. A solution that is politically acceptable may not be good for either the environment or the economy.

---

[255]T. M. L. Wigley, "The Kyoto Protocol: CO_2 CH_4 and Climate Implications, *Geophysical Research Letters*, July 1, 1998, pp. 2285–88.

[256]Bjørn Lomborg, *The Skeptical Environmentalist*, p. 304.

[257]Stephen Castle, "EU Sends Strong Warning to Bush Over Greenhouse Gas Emissions," *The Independent*, March 19, 2001, p. 14.

[258]Steven Hayward and Julie Majeres, *Index of Leading Environmental Indicators*, 6th ed. (San Francisco: Pacific Research Institute for Public Policy, 2001), p. 61.

[259]Brink Lindsey, *Against the Dead Hand*, p. 253.

Another proposed government-based solution is the imposition of a carbon tax. The purpose of such a tax would be three-fold:

1. Discourage the use of carbon-based fuels.
2. Encourage the use of alternative energy sources (both by raising the price of conventional fuels, and by providing government with the means to subsidize alternatives).
3. Provide funds for government energy research.

Some economists argue that a carbon tax would be more flexible and straightforward than Kyoto's cap-and-trade scheme. The main objections to such taxes are that they would drive consumers away from more efficient sources of fuel and towards less efficient (though government-approved) sources. Also, a tax of any kind shifts financial resources to the government and away from a private sector better equipped to develop new energy technology.

Most scientists have concentrated on controlling carbon dioxide as the solution to global warming. However, Richard Lindzen and James Hansen (two prominent scientists who are usually on opposite sides of the debate) have argued that a better approach, at least in the near term, may be to focus on more powerful, and more easily controlled, warming agents such as methane, soot, and chlorofluorocarbons.[260]

Carbon sequestration may also be a more viable method of reducing atmospheric CO_2 concentrations than reducing carbon emissions. A study published in *Science* found that large quantities of carbon could be captured at an estimated cost of $30 a ton, or about $13 per barrel of oil or $0.25 per gallon of gasoline. The author, Klaus Lackner, concluded, "Today's urgent need for substantive CO_2 emission reductions could be satisfied more cheaply by available sequestration technology than by an immediate transition to nuclear, wind, or solar energy. Further development of sequestration would assure plentiful, low-cost energy for the century, giving better alternatives ample time to mature."[261]

On the other hand, while the International Energy Agency agrees that "carbon sequestration and storage technologies hold out the long-term prospect of enabling fossil fuels to be burned without emitting carbon into the atmosphere," it cautions that "these technologies . . . are unlikely to be deployed on a large scale before 2030."[262]

The issue of population control often arises in discussions of global warming. By limiting the number of people, it is argued, resource consumption and

[260]Andrew Revkin, "Study Proposes New Strategy to Stem Global Warming," *New York Times*, August 19, 2000, p. A12; Richard Lindzen, "Scientists' Report Doesn't Support the Kyoto Treaty," *Wall Street Journal*, p. A22.

[261]Klaus Lackner, "A Guide to CO_2 Sequestration," *Science*, June 13, 2003, p. 1678.

[262]International Energy Agency, *World Energy Outlook: 2002* (Paris: OECD/IEA, 2002), p. 31.

pollution will also be limited. It turns out, however, that population growth is already slowing for economic reasons. Contrary to what Thomas Malthus believed, people do not "breed like flies." Instead, in the case of procreation, as in other areas, people tend to act rationally given their circumstances.

In western countries, population growth has slowed or even halted because children are a net economic burden. In a high-tech society, children are not suitable as sources of labor but must be educated for many years before they become productive. By that time they are ready to leave home and start households of their own.

In developing countries, by contrast, little education is needed before children become proficient with the lower levels of technology available. Children in these countries are considered cheap labor and an economic asset. In addition, children provide for their elderly parents. Finally, child mortality rates are high in the Third World, and people tend to have additional children to ensure that at least some will survive to add to the family's income. Attempts by the West to reduce population growth in these countries are often resented by locals as attacks on their wealth.

As the Third World nations advance, however, people there will change their actions as incentives change. In fact, the world population growth rate has been declining since about 1970.[263]

Julian Simon's contrary view was that, "In the long run the most important economic effect of population size and growth is the contribution of additional people to our stock of useful knowledge. And this contribution is large enough in the long run to overcome all the costs of population growth."[264]

Whether a person is seen as a mind and pair of hands or as just another mouth to feed may depend on the society in which he or she lives. People in free-market societies are rewarded in proportion to what they produce, and, as a result, they produce far more than they consume. The guiding principle of socialism, on the other hand, is "from each according to his ability, to each according to his need." To the extent that this rule is actually followed, it creates incentives for people to demonstrate minimum ability and maximum need.

Some rather exotic methods of dealing with global warming have also been proposed. For example, the oceans near the equator could be fertilized with iron filings to promote the growth of plankton. When these billions of tiny plants die, they sink to the bottom of the ocean. There is little or no oxygen at the ocean floor, and plankton do not decay. The carbon that they took out of the atmosphere and incorporated in their bodies while they were alive remains locked up.

---

[263]Bureau of the Census, U.S. Department of Commerce, http://www.census.gov/ipc/www/worldpop.html

[264]Julian Simon, *The Ultimate Resource*, p. 196.

A problem with this idea is that mining and refining iron ore requires a lot of energy, and still more energy would be spent flying over the ocean to drop it. It is questionable whether the additional plankton engendered by this method would offset the CO_2 emitted in the production and delivery of the iron fertilizer.

> When assessing the value of any proposal, it is important to look at the big picture (what economists call performing a *life-cycle analysis*). For instance, the federal government has proposed new efficiency standards for household appliances that could help reduce energy requirements (and therefore CO_2 emissions). This may seem to be a good idea at first glance. But will the energy savings over the life of an appliance offset the additional energy and resources needed to make it more efficient? Or even if there is a net benefit, could the resources expended on improving appliance efficiencies be spent to greater effect in other ways?

Scientists at Los Alamos National Laboratory have proposed what they claim is a cost-effective method for removing carbon dioxide directly from the atmosphere. Their process involves passing air over calcium oxide (quicklime), which combines with carbon dioxide in the air to form calcium carbonate (limestone). The limestone is then heated to yield pure CO_2 and quicklime. The quicklime is recycled back to the extractor, while the CO_2 can be injected into the ground.

"Geoengineering" techniques designed to block sunlight from entering the Earth's atmosphere may also offer a partial solution. As explained in an article in *Science*, such measures might include placing "layers of reflective sulfate aerosols in the upper atmosphere. . ., injecting sub-micrometer dust into the stratosphere. . ., increasing cloud cover by seeding," and placing huge (2000 km diameter) mirrors in space to block the sun's radiation. [265]

The possibility of such future technology leads to an important question: Should we attempt to mitigate global warming now or wait until we understand the potential problem more clearly and have better (and as yet unimagined) technology to handle it?

> NASA has offered what is, perhaps, the most novel solution to global warming yet proposed—move the Earth to a higher orbit, farther from the sun. According to the plan's authors, Greg Laughlin, Don Korycansky, and Fred Adams of NASA's Ames Research Center, it is all basic physics. Simply locate a suitable asteroid, attach a rocket to it, fire the rocket at just the right time to alter its course, and sling it by the Earth so that its gravitational force drags our planet into a higher orbit.

---

[265]Martin Hoffert, et al., "Advanced Technology Paths to Global Climate Stability," p. 986.

> There are several possible downsides to their plan, however. The first is that a miscalculation could send the asteroid crashing into the Earth. Another is that if the plan works, the moon could be left behind. Also, with the Earth at a higher orbit, a year (the time it takes the planet to travel around the sun) would be longer. The loss of the moon and a longer year would each have significant impacts on global climate.
>
> Actually, the scientists proposed the plan as a solution to the ultimate global warming problem. Our sun is gradually brightening, and in about a billion years it will be hot enough to kill off all life on Earth if the planet stays in its current orbit.[266]

TRADE-OFFS

Normally, the most environmentally friendly technology is also the most efficient, and people will tend to move to such technologies of their own accord. Yet this is not always the case. While it is clearly much more efficient to strip mine coal by simply removing the overburden and digging out the coal, this technique leads to erosion that fills rivers and lakes in the area with silt and heavy metals. Government regulations, therefore, now require coal companies to scrape off and preserve the topsoil before the rest of the overburden is removed. After the coal has been extracted, the overburden is replaced and contoured. Next, the topsoil is restored and seeded to prevent erosion.

The benefit is that the land is preserved, as are lakes and rivers. At the same time, though, more resources must be expended to extract the same amount of coal. The result is an environmental trade off—one thing is given up to gain another.

Nowadays, coal ash need no longer be simply carted away and dumped. Instead it can be used in cement making. However, the ash can be used for this purpose only if its carbon content is very low; in other words, the coal must be thoroughly burned. Such complete burning requires higher temperatures, which increases power plant efficiency but also produces more nitrogen oxide emissions.

Such trade-offs can be wrenching. Many environmentalists have worked hard to save natural wetlands, yet wetlands are significant sources of methane, a greenhouse gas.

Similarly, support is growing for the practice of organic farming, which eliminates the use of pesticides and fertilizers. Banishing these chemicals from agriculture, however, would reduce the amount of food that can be grown on an acre of land. In order to switch to organic farming, then, more land must be put to the plow. This means cutting down forests that remove carbon from the atmosphere.

---

[266]Frederic Golden, "How Long Will We Be Around?," *Time*, June 25, 2001, p. 53.

European environmentalists generally suport the diesel engine as a viable alternative to the gasoline engine because of its greater efficiency and lower greenhouse gas emissions. Several diesel passenger-vehicle models already on the road get more than 60 miles per gallon. Moreover, the infrastructure needed to supply diesel fuel to consumers is already in place. Many environmentalists in America, however, oppose a switch to diesel engines because they emit more NO_x and soot particles than do gasoline engines. Soot in the atmosphere has a net warming effect, and may more than offset any gains from reduced CO_2 emissions.

WHAT ABOUT POORER COUNTRIES?

Some have suggested that, because the industrial nations produce most of the anthropogenic CO_2, they have a moral obligation to help the world's poorer countries cope with the impact of enhanced global warming. Others reject this notion because, they argue, Asian rice paddies produce massive amounts of methane, a more powerful greenhouse gas than carbon dioxide. Also, they point out that the large populations in China and India are going to produce a tremendous amount of CO_2 in the future as they become more technologically advanced and their energy needs increase.[267] Such finger pointing is one reason that treaty negotiations are so difficult.

In any event, if the West chooses to help the poorer nations, it must decide how best to do it. In the past, foreign aid has usually come in the form of government-to-government payments or loans. All too often, the money has disappeared into useless public works or into private, offshore bank accounts. Worse, the money may prop up corrupt and repressive regimes that destroy people's freedom and confiscate the products of their labor.

The wealthier people are, the more they will be able to command the resources needed to deal with a warmer climate and rising sea levels. History teaches that the freer a people are, the wealthier they are. Rather than strengthening repressive governments, then, perhaps private property rights and open markets should be strengthened.

BUILT-IN BIASES

Understandably, global warming has become a very emotional issue for many people, and these emotions can make the job of finding the truth difficult. Further clouding the facts are the built-in biases that everyone has. Sometimes

---

[267]The U.S. Energy Information Administration estimates that combined China/India emissions will exceed those of the U.S. by the year 2020 [EIA, *International Energy Outlook* 2002 (Washington: Department of Energy, 2001), Table A10.

these biases depend upon personality types such as whether an individual is generally an optimist or a pessimist.

People's livelihoods also influence their attitudes. Neither oil producers nor coal miners want their products banned or restricted, and they would prefer that the global warming issue simply went away. On the other hand, if it turns out that global warming is not a problem, then many researchers will lose their government funding. Professional environmentalists need hot issues to garner contributions and keep themselves relevant. Journalists welcome global warming as a front-page issue on slow news days. They like crisis reporting and the bold headlines that go with it; negative stories sell more newspapers than do positive ones. People in government tend to like crises as well. Governments increase their power by offering solutions to problems, both real and imagined. The bigger the problem, the more people look to government institutions for answers.

That is not to say that everyone with a stake in the issue will consciously try to hide the truth or skew the data. But people often emphasize facts that support their own positions and either ignore or minimize information to the contrary.

Whether it is decided that global warming is or is not a problem, the decision must be based on good science and economics, and not emotion. Actions must be well thought out; good intentions are not enough. Too many people can be hurt if bad policies are adopted. As even energy critic Paul Ehrlich said, "Taking action on the basis of worst-case prognoses would . . . be inappropriate and costly; suddenly imposing fuel rationing and high taxes on industrial activity with no tangible justification would cause economic disruption and most likely would backfire."[268]

❝❝Action is most effective when it is driven by passion but directed by reason.**❞❞**

The authors

WHERE DOES ALL THIS LEAVE US?

Thus far, neither the computer models nor the actual data have provided a clear and definitive picture of the trends in the Earth's climate. Scientists are still arguing about how serious a problem enhanced global warming is and even if there is a *greenhouse signal* apart from natural variability. If the experts cannot agree, how are laymen to decide the truth? The fact is that we just do not know yet what the truth is, and much more research is needed before we can know.

In the meantime, some environmentalists argue that the *precautionary principle* requires that we act now even though we lack scientific certainty as to

---

[268]Paul Ehrlich and Anne Ehrlich, *Betrayal of Science and Reason*, p. 31.

whether, or to what degree, human activity has enhanced global warming, how much it will affect future climate, or, more importantly, what impact it will have.[269] The precautionary principle has been defined in a number of ways, ranging from Principle 15 of the United Nations' Rio Declaration (1992):

> In order to protect the environment, the precautionary approach shall be widely applied by States according to their capabilities. Where there are threats of serious or irreversible damage, lack of full scientific certainty shall not be used as a reason for postponing cost-effective measures to prevent environmental degradation.

to the more radical Wingspread Declaration (1999):

> When an activity raises threats of harm to human health or the environment, precautionary measures should be taken even if some cause and effect relationships are not established scientifically. In this context the proponent of the activity, rather than the public, should bear the burden of proof.

Critics point out that the principle must be applied in a balanced way; that is, it should be applied not only to the activity that environmentalists want regulated, but also to the regulations themselves. Government intervention can make a problem worse[270] or can create problems where none existed before. As pointed out in a study by the National Academy of Sciences, "Errors of doing too much can be as consequential as errors of doing too little; the error of trying to solve the wrong problem is as likely as the error of failing to act."[271]

Indur Goklany, an American representative to the IPCC and chief of the Technical Assessment Division of the National Commission on Air Quality, has proposed the following climate-change policies as being consistent with a balanced application of the precautionary principle.[272] Note that Goklany takes into consideration the impact on both public health and the environment.

1. Avoid government mandated greenhouse gas emission restrictions in the next few decades. In the absence of such mandates, individuals, businesses and other institutions would, in any case, undertake no-regret actions which, by definition, would pay for themselves even

---

[269]Indur Goklany, "Potential Consequences of Increasing Atmospheric CO_2 Concentration Compared to Other Environmental Problems," *Technology*, vol. 7, Supplement (2000), pp. 189–213.

[270]Remember President Jimmy Carter's *Powerplant and Industrial Fuel Use Act of 1978*, which was intended to save natural gas, but resulted in increased coal power plant construction and with it more pollution and CO_2 emissions.

[271]National Academy of Sciences, et al., *Policy Implications of Greenhouse Warming*, p. 194.

[272]Indur Goklany, *The Precautionary Principle: A Critical Appraisal of Environmental Risk Assessment* (Washington: Cato Institute, 2001), pp. 86–88.

in the absence of any climate change. Mandating controls that would necessarily have to go beyond no-regret actions to be meaningful, will likely slow worldwide economic growth leading to more hunger, worse health, and higher mortality rates, especially in the Third World.

2. Avoid artificially raising oil and gas prices. Higher prices would slow the switch from dirtier fuels such as wood, coal, and animal dung. Higher energy costs would also increase the costs of food production and reduce crop yields. Lower crop yields would, in turn, lead to increased land conversion and loss of habitat, which, in turn, would only add to CO_2 emissions.

3. Work to solve problems, such as malaria and malnutrition, which are urgent today and may be aggravated by global warming.

4. Eliminate policies (such as subsidies for the exploitation of energy and other natural resources) that contribute to increased production of greenhouse gases.

5. Increase agricultural productivity to expand food production while at the same time reducing the amount of land under cultivation. This would increase natural habitat and decrease soil erosion and the associated loss of carbon sinks.

6. Increase people's ability to adapt to environmental change by promoting technological progress, trade, and economic growth through the institutions of free markets, secure property rights, and honest government.

7. Continue researching the science and economics of climate change. Such research should include preventative and corrective measures for dealing with greenhouse gases and global warming as well as ways to adapt to a changing climate in the event that there are problems.

Goklany's first point—the avoidance of near-term government-mandated emission reductions—is very important. As discussed, many ways of dealing with greenhouse gases have already been proposed. Many other methods will be developed in the future as we learn more. Money spent on ineffective solutions now cannot be spent on things that will actually make a difference. Worse, measures that harm the economy now will reduce the resources that will be available in the future when there will be a better understanding of the problem and of how to deal with it.

Unfortunately, most government initiatives have zeroed in on one very expensive and very ineffective solution—compulsory CO_2 emission reductions. The Kyoto Treaty and various U.S. House and Senate proposals would impose caps on CO_2 emissions at very high cost for virtually no benefit in terms of reductions in the impacts of climate change or in terms of advancing human or environmental welfare.[273]

If anthropogenic global warming proves to be a problem, then we must keep our eyes on the goal—the reduction or reversal of the effects of such warming. To achieve this goal efficiently and effectively in the long run, we should examine a combination of measures that would (a) reduce emissions to slow down temperature change, (b) remove greenhouse gases from the atmosphere, and (c) help societies cope with the negative impacts of climate change.

While establishing this goal may be a legitimate function of government, it will be counterproductive for government to dictate a one-size-fits-all solution. Reducing average global temperature will be a titanic undertaking, and, if it must be done, it must be done as efficiently as possible. Trying to handle such a vast challenge inefficiently will generate pollution, waste resources, perpetuate poverty, and engender public anger.

Only market forces can marshal the incredible creativity needed to tackle such an undertaking. The solutions may include planting trees, changing farming methods, managing forests differently, using alternative fuels, improving emissions controls, and employing technology that we cannot even imagine today. Whatever the answers, they can only come from unshackled, inventive minds and from a dynamic marketplace, free to employ resources to their best effect.

> **❝***Perhaps the most important aspect of [global warming] . . . is that we now have ever-increasing capacities to reverse such trends if necessary. And we can do so at costs that are manageable rather than being an insuperable constraint upon growth or an ultimate limit upon the increase of productive output or of population.***❞**[274]
>
> *Julian Simon—American professor of business administration*

---

[273]Goklany, *The Precautionary Principle*, p. 67.

[274]Julian Simon, "Introduction," in Simon, ed., *The State of Humanity* (Cambridge, MA: Blackwell, 1995), p. 18.

ENERGY FOR THE FUTURE

7

LOOKING FOR TROUBLE

The year before his death, Julian Simon made what he called his "long run forecast . . . in brief:"

> The material conditions of life will continue to get better for most people, in most countries, most of the time, indefinitely. Within a century or two, all nations and most of humanity will be at or above today's Western living standards.
>
> I also speculate, however, that many people will continue to *think* and *say* that the conditions of life are getting *worse*.[275]

Enormous material progress has been made in the past two hundred years. Much of this progress was the result of advances in energy technology made by people living in freedom. Moreover, these advances are accelerating even as the environment, at least in the West, improves.

One would think that energy alarmism, along with its unbroken string of failed predictions and policies, would have been buried under this mountain of good news long ago. Yet today, alarmism is alive, well, and newsworthy. Scary computer-generated climate-change scenarios, rising natural gas and gasoline prices, and turmoil in the Middle East have made energy pessimism nearly as popular as it was during the American energy crises of the 1970s and the British coal panic of the 1860s.

Examples abound. The *New York Times* reports that Saudi Arabia's oil fields are "tired."[276] One popular book predicts that global oil production will reach

---

[275]Julian Simon, quoted in Ed Regis, "The environment is going to hell . . ." *Wired*, February 1997, p. 198. Bjørn Lomborg, who discovered Julian Simon by reading this article, used Simon's prediction as the epigraph for *The Skeptical Environmentalist*.

[276]Jeff Gerth, "Forecast of Rising Oil Demand Challenges Tired Saudi Fields," *New York Times*, February 24, 2004, p. A1.

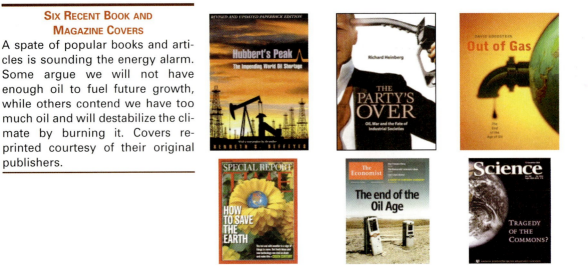

its "Hubbert's Peak" by 2009. Others declare that the "fabulous party" of
cheap energy is over, "civilization as we know it will come to an end sometime in this century unless we can find a way to live without fossil fuels," and
crude oil will reach $100 a barrel by decade-end.[277]

Sometimes, these alarms are contradictory. On the one hand, we are
told the world will soon run out of carbon-based fuels, while on the other
hand predictions of global warming Armageddon rely on the assumption
that the world will continue to burn more and more of these fuels for
decades to come.

The statistical record of improvement over the last two centuries challenges these gloomy scenarios and the pessimism behind them. By any
index—availability, affordability, reliability, cleanliness, efficiency, utility—
the long-term energy trends have been positive. Problems have been faced and
solved by creative people, and, in special cases (such as air pollution), incre-
mental regulation. Our own predictions, based in part on official forecasts, reflect those trends.

---

[277]Kenneth Deffeyes, *Hubbert's Peak: The Impending World Oil Shortage* (Princeton: Princeton University Press, 2001), p. 158; Richard Heinberg, *The Party's Over: Oil, War and the Fate of Industrial Societies* (Canada: New Society Publishers, 2003), p. 242; David Goodstein, *Out of Gas: The End of the
Age of Oil* (New York: W. W. Norton, 2004), p. 15; Stephen Leeb and Donna Leeb, *The Oil Factor:
Protect Yourself—AND PROFIT—from the Coming Energy Crisis* (New York: Time Warner, 2004), p. 50.

TWO POSITIVE TREND BOOKS

The statistical record of the last century shows long-term trends of improving human welfare. Short-term trends can be negative (e.g., rising energy prices since 2000), but they spur adjustments that shape longer-term positive trends. Covers reprinted courtesy of their original publishers.

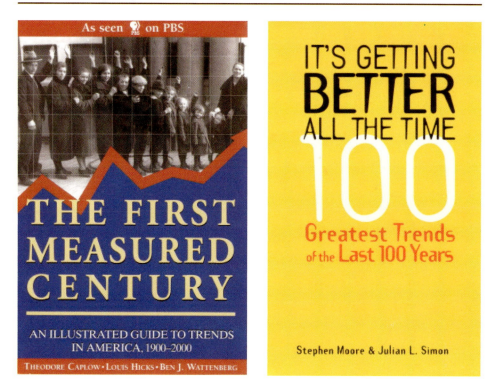

There is, however, one urgent energy alarm that must be sounded. Not a hypothetical warning of something that may happen in the distant future, but a major energy sustainability problem that is happening right now. It concerns one-fourth of the world's inhabitants. It is *wretched energy poverty*, the lack of electricity and fuel for heating and transportation and of all that modern energy provides: clean air and water, adequate lighting, medical facilities, education aids, communication infrastructure, labor-saving devices, and more.

This energy sustainability problem is not the result of fossil-fuel depletion or combustion. It is a child of *statism* and all the philosophies behind it—philosophies that justify coercive government control of economies, of people's lives, and of the energy resources that make life possible.

Major reforms are needed to eradicate—house by house, village by village—the poverty that comes from a lack of basic market institutions such as property rights, property titles, and the freedom to exchange, contract, and associate with others.

LOOKING AHEAD

The perils of prediction are well known. An earlier chapter of this book quotes a number of people whose pessimistic forecasts have proved to be spectacularly wrong. There are many pitfalls waiting for those who venture to peer into a crystal ball.[278]

First, people often assume, as Thomas Malthus did more than two hundred years ago, that technology will not change significantly and new substitutes will not emerge. If they do foresee changes, they often see improvement in some favored areas (e.g., solar and wind power), while ignoring the fact that the other technologies will also be improving at the same time. Alternative fuels will not compete with conventional fuels as they are but as they will *become*.

Some forecasters tend to over-estimate the changes in the near term (20 years out), and underestimate the changes in the long term (50 to 100 years). Near-term overestimation occurs because forecasters often do not take into account the tremendous inertia in any system.

Some proponents of energy transformation predict that hydrogen will replace gasoline as the primary transportation fuel in the next few decades. But even if all the technical problems with using hydrogen were to be resolved, it would take years for a hydrogen infrastructure to be established and for the vehicles now on the road to be replaced. Large hydrogen generation units would have to be constructed, along with the electric plants to provide for their power needs. In addition, hydrogen transportation and delivery facilities would have to be created. All this would take enormous effort and capital, and none of it would happen overnight. In the meantime, millions of new vehicles with a life expectancy of at least 10 years—virtually all fueled by gasoline or diesel—are added to the fleet every year.

Long-term underestimation of future technology stems from our inability to imagine the directions that even existing technology can take during large time spans. As late as the 1970s, many experts were confidently predicting that con-

---

[278]See Ch. 4, pp. 81–82.

sumption of electricity would level off in the United States because most homes already had nearly all of the household appliances of the day. Few could predict the explosion in the use of personal computers, the creation of the Internet, and the resulting rise in demand for electricity.[279] Even today, many new uses for electricity are appearing that consumers welcome.

Moreover, we are unable to predict the breakthroughs in basic science that will be made. Who in 1900 could have foreseen the mapping of the human genome, nanotechnology, or superconductivity? Who today can imagine the impact these advances will have on society in fifty or a hundred years? What we do know is that these discoveries will open new horizons for new discoveries and breakthroughs.

> *The future is inevitably and irreducibly unpredictable, for the simple reason that we cannot now know what still remains to be known.*[280]
>
> **Brink Lindsey**

> *The outstanding fact of history is that it is a succession of events that nobody anticipated before they occurred.*[281]
>
> **Ludwig von Mises**

Another problem in forecasting is the inability to differentiate between what is *possible* and what is *practical*. As technologist Mark Mills points out, "It is possible to build personal, Buck Rogers-style jet backpacks. They were first built in the late 1950s. They're commercially available today. They're just not useful."[282] The creation of large solar farms is possible, but such farms have not proved practical.

Keeping in mind these pitfalls, it is with great caution that we venture to present a summary of forecasts from well-known energy agencies and to make a few predictions (qualitative, not quantitative) of our own.

---

[279]We (the authors of this book) were reminded of the difficulty in identifying new technologies that will have a lasting impact on society as we were selecting items for the timeline in *Appendix A*. It is a safe bet that the Wright Brothers were really onto something with their flying machine, but how does one identify inventions made in the last decade or two that will have the same sort of lasting impact?

[280]Brink Lindsey, *Against the Dead Hand*, p. 51.

[281]Ludwig von Mises, *Theory and History* (Arlington, VA: Arlington House, 1969), p. 378.

[282]Mark Mills, *Getting It Wrong: Energy Forecasting and the End-of-Technology Mindset*, p. 30.

SHORT-TERM FORECAST

The short-term outlook for energy, as for other goods in a market economy, is that it will remain much as it is today. Problems will arise and will be resolved, leaving us better off than before. This is an easy prediction to make because free markets inspire solutions to problems.

Assume, for example, that the price jumps for a particular fuel. Abnormally high prices will spur buyers and sellers into remedial action. High prices will lead consumers to conserve and to look for substitutes. At the same time, high prices will encourage suppliers to find and produce more of the fuel in order to increase their profits.

For instance, after a decade in which the delivered price of natural gas to American power plants was under $3 per million BTU (MMBTU), prices since late 2000 have averaged close to $5 per MMBTU. This 60 percent sustained increase cannot be blamed on a sudden scarcity in the natural resource itself given that proved gas reserves are the highest level ever recorded in both the United States and Canada. Rather, demand for this cleanest of carbon-based fuels simply grew faster than did the infrastructure that supplies it.

In response, producers are drilling new gas wells at a brisk pace. Suppliers are applying to the Federal Energy Regulatory Commission (FERC) for permission to import more liquefied natural gas from abroad and to build new terminals to receive the LNG. Several proposals to pipe natural gas from Canada's Northwest Territories and Alaska to the Lower-48 states are under consideration by investors and regulators. New gas storage facilities are being built to take advantage of price volatility (buy low, store, sell high). Meanwhile, electric utilities are turning, where possible, to substitute-energies to generate the power they need. Finally, higher gas prices are promoting conservation.

U.S. Coast Guard

Movement is also taking place on the political front. Gas pipeline projects in the United States that have been delayed by regulatory roadblocks are being revisited. Promising geologic structures that are on public lands and that have been politically off-limits are getting another look. The Rocky Mountain area and areas off the East, West, and Florida coasts are of particular interest.

Given all this activity, it might be a good bet that the price of natural gas will be lower in the next few years than it has been in 2001–2004. In fact, that is exactly how commodity traders *are* betting as indicated by futures prices on the New York Mercantile Exchange.

Suppose, on the other hand, that the price of a fuel falls below its historical average. Then a bet on a longer-term price rise might be a good one. This is how markets work in the continually unfolding short term.

Much of the cyclical nature of prices is due to imperfect knowledge. Producers do not know what future demand for their products will be, though they try to forecast demand as best they can because the rewards for getting it right can be high, and the penalties for getting it wrong can be severe.

Take the case of a natural gas company that correctly predicts higher future demand and expands its facilities in time to take advantage of that demand. Due to its foresight, the company stands to increase its profits and gain market share. If, on the other hand, the predicted increase in demand does not materialize, the expanded facilities could lie idle and the company might lose its investment. If the firm goes ahead and produces more fuel anyway, the additional supply will depress prices (though such lower prices will, eventually, result in higher demand).

Finally, if the company fails to foresee an increase in demand, it may well lose business to competitors whose vision was clearer. While the company may try to quickly increase its production capacity, it may be years before it can get the necessary government permits and complete the construction work.

Fuel consumers have similar incentives to correctly anticipate future prices. Suppose a power company plans to build a new electricity generation plant. It wants to use the cheapest fuel available. If the price of natural gas is currently high, should it build a coal-fired plant? But by the time the plant is built, natural gas producers may well have caught up with demand, and the price of their product could again be competitive. Typically, developers hedge their bets by entering into long-term contracts to purchase fuel and to sell power.

This corrective process never ends. The market is not perfect and problems occur, but problems are also a key driver of progress. We learn by trial and error, and progress happens in fits and starts, but it does happen.

MID-TERM U.S. FORECAST

The forecasting arm of the U.S. Department of Energy, the Energy Information Administration (EIA), has estimated American energy supply and demand through the year 2025. Based on a projected economic growth of 3 percent a year, EIA sees energy demand increasing by 1.5 percent a year, indicating an increase in energy efficiency, or (stated another way) a decrease in *energy intensity*, of 1.5 percent per annum (see next page).

Use of all three carbon-based fuels is forecast to increase significantly during the forecast period. The EIA estimates that non-hydropower renewables (ethanol, geothermal, biomass, solar, and wind) will provide less than seven percent of America's total energy by the year 2025.[283] However, because alternative energies largely depend upon government subsidies, this figure will shift if government support changes one way or another.

MID-TERM GLOBAL FORECAST

The EIA has also made a forecast of world energy supply and demand through 2025 (see page 184). It sees total demand rising by almost 60 percent—a growth rate of nearly 2 percent per year.[284] These projections assume an annual decline in energy intensity of 1.1 percent.

World demand is expected to increase faster than in the United States as the developing world strives to catch up with the West. As it has in the past, electricity demand is expected to grow at a faster pace than overall energy demand.

Oil use will rise along with the transportation market as gasoline and diesel will remain the dominant fuels. Natural gas is expected to be the fuel of choice for electricity generation because of its superior economics and relatively low environmental impact. Coal consumption is also expected to increase, but at less than the overall average. In fact, during the forecast period, natural gas use is predicted to overtake coal for the first time in history. Nuclear power is expected to remain flat, and renewables, still dominated by hydroelectricity, will grow near the average.

The combined market share of carbon fuels will increase to 88 percent from 85 percent in the forecast period. More than 90 percent of the increase in total energy demand is expected to be met by oil, gas, or coal.

---

[283]U.S. Energy Information Administration, *Annual Energy Outlook* 2003 (Washington: Department of Energy, 2003), tables A1 and A18.

[284]U.S. Energy Information Administration, *International Energy Outlook* 2003 (Washington: Department of Energy, 2003), pp. 183, 188, 190.

Forecast of U.S. Energy Demand

Robust energy demand growth will be met by all the carbon-based energies—oil, gas, and coal. Renewable energy growth is higher than the overall average but still small in absolute terms. *Source:* U.S. Energy Information Administration, *Annual Energy Outlook 2004,* p. 133.

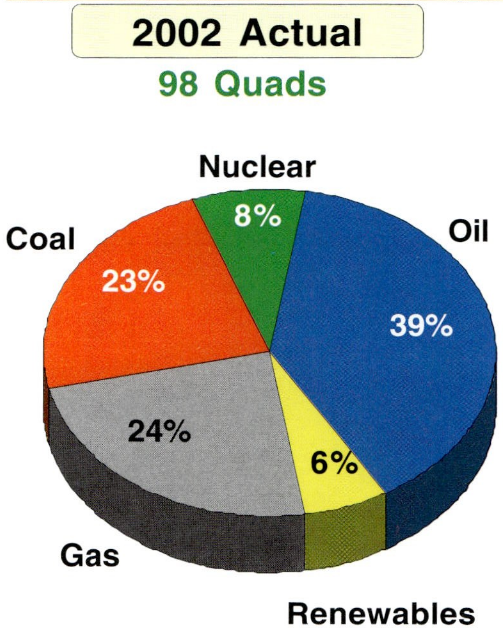

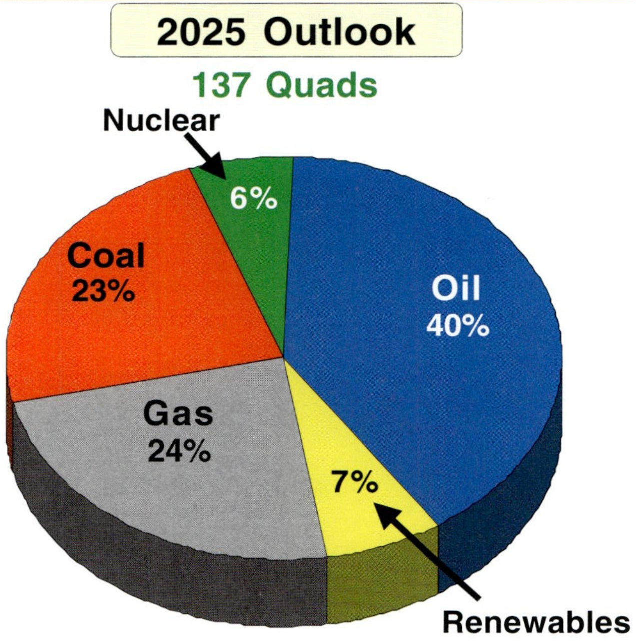

2002 Actual
98 Quads
Nuclear
8%
Coal
23%
Oil
39%
24%
6%
Gas
Renewables

2025 Outlook
137 Quads
Nuclear
6%
Coal
23%
Oil
40%
Gas
24%
7%
Renewables

Quadrillion Btu	2002	2025	Change	Growth	Growth/yr.
Coal	22	32	10	43%	1.6%
Gas	23	32	9	38%	1.4%
Nuclear	8	9	1	5%	0.2%
Renewables	6	9	3	54%	1.9%
Oil	38	55	17	44%	1.6%
Total	98	137	40	40%	1.5%

Demand for natural gas is forecast to grow the most in the coming decades, but coal and oil usage will remain strong. Renewables have the highest growth rate but start from a low base, so their overall contribution will remain modest. *Source:* U.S. Energy Information Administration *International Energy Outlook 2003* p. 183.

2001 Actual

404 quads

2025 Outlook

640 quads

Quadrillion Btu	2001	2025	Change	Growth	Growth/yr.
Coal	96	139	43	45%	1.6%
Gas	93	182	89	96%	2.8%
Nuclear	26	29	3	12%	0.3%
Renewables	32	50	18	56%	1.9%
Oil	157	241	84	54%	1.8%
Total	404	640	237	59%	1.9%

LONG-TERM FORECAST

Oil, natural gas, and coal will remain abundant throughout the twenty-first century at prices competitive with other energy alternatives.

While some argue that world oil production will peak in the next one to two decades, the mainstream view is more optimistic. The World Energy Council stated, "The conclusion can be reliably drawn that fossil fuel resources are adequate to meet a wide range of possible scenarios through to 2050 . . . and well beyond."[285] The Intergovernmental Panel on Climate Change found that "there are abundant fossil fuel resources that will not limit carbon emissions during the 21st century."[286] In fact, the IPCC estimates that only about 1.5 percent of the total potential hydrocarbons in the Earth's crust has been consumed.[287]

This does not mean that energy sustainability issues have been put to rest. As explained in Chapter 6, the question is not "will we will run out of fossil fuels?" but "what happens to the environment if we burn what we have?" Harvard environmentalist John Holdren questions whether carbon-based fuels will remain affordable if the cost of correcting the environmental damage they do is included in their price.[288]

Corbis

[285]World Energy Congress, *Living in One World: Sustainability from an Energy Perspective* (London: WEC, 2001), p. 161.

[286]IPCC, *Climate Change 2001: Mitigation*, p. 4.

[287]Ibid., p. 236.

[288]In Holdren's words, the energy sustainability problem stems from "environmental impacts and sociopolitical risks—and, potentially, of rising monetary costs for energy when its environmental and sociopolitical hazards are adequately internalized and insured against." John Holdren, "Energy: Asking the Wrong Question," *Scientific American*, January 2002, p. 65.

However, so far no alternatives can compete with the convenience, portability, efficiency, or cost of oil, gas, and coal. In addition, carbon-based fuels will continue to be made cleaner and more efficient.

As the following chart shows, the renewable energy era has already come and gone. Even though alternative energy sources like biomass, alcohol, and solar and wind power are being touted as energy sources for the future, they are actually throwbacks to the past.

Eventually, however, fossil fuels will be replaced with other primary energy sources. A new form of energy will ease out hydrocarbons when consumers judge it a better product at a better price. If the transition is left to the free market, it will happen in one of two ways:

1. If hydrocarbons become increasingly scarce, their cost will increase until substitutes are found, just as kerosene once replaced whale oil.

2. A technological breakthrough, or cluster of breakthroughs, will make some substitute more efficient and less costly.

U.S. ENERGY CONSUMPTION: 1775-2000

During the past 225 years, the United States has experienced two distinct energy eras: the renewable era and the (current) carbon-fuel era. *Source:* U.S. Energy Information Administration, *Annual Energy Review 2001,* pp. 355–357.

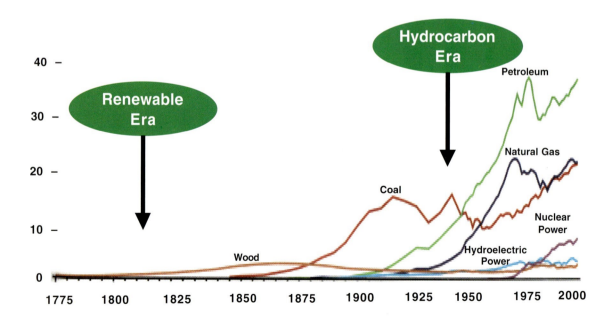

❝[E]nergy can be produced indefinitely so long as human ingenuity is allowed to keep up with demand. Any such perceived shortage is simply a practical problem to which man's creativity will, if permitted, find a solution. . . . This is the lesson of history, and the reason we need not fear any permanent energy shortage. Human ingenuity carried us from firewood to coal, from whale oil to petroleum. And human ingenuity will carry us to the next stage of energy evolution.**❞**[289]

Rabbi Daniel Lapin

❝It's reasonable to expect the supply of energy to continue becoming more available and less scarce, forever.**❞**[290]

Julian Simon

While the prices of individual fuels may rise, there is little reason to believe that energy *per se* will grow less abundant and more costly. The lesson of history is that in free societies individuals produce more energy than they consume. As Julian Simon wrote:

> There is no physical or economic reason why human resourcefulness and enterprise cannot forever continue to respond to impending shortages and existing problems with new expedients that, after an adjustment period, leave us better off than before the problem arose. Adding more people will cause us more such problems, but at the same time there will be more people to solve these problems and leave us with the bonus of lower costs and less scarcity in the long run. The bonus applies to such desirable resources as better health, more wilderness, cheaper energy, and a cleaner environment.[291]

ENERGY AND POVERTY

According to the International Energy Agency, "some 2.4 billion people rely on traditional biomass—wood, agricultural residues and dung—for cooking and heating."[292] And typically this biomass is burned very inefficiently with much of the heat lost. People in developing nations often have no incentive to use such fuel efficiently because it costs them nothing more than the effort to gather it.

Yet, there are unseen costs to using these so-called free fuels. Deforestation is a serious problem in many parts of the world. Because no one owns the forests and jungles, no one has an incentive to conserve them—the tragedy

---

[289]Rabbi Daniel Lapin, "Existential (Energy) Crisis," *National Review Online*, June 11, 2001.

[290]Julian Simon, *The Ultimate Resource* 2 (Princeton: Princeton University Press, 1996), p. 181.

[291]Ibid., p. 580.

[292]International Energy Agency, *World Energy Outlook: 2002*, p. 365.

GLOBAL ENERGY POVERTY

The energy poor: people with no access to electricity and those with some electricity but who must still rely on primitive biomass for heating and cooking. *Source*: International Energy Agency, *World Energy Outlook 2002*, p. 372. Copyright OECD/IEA.

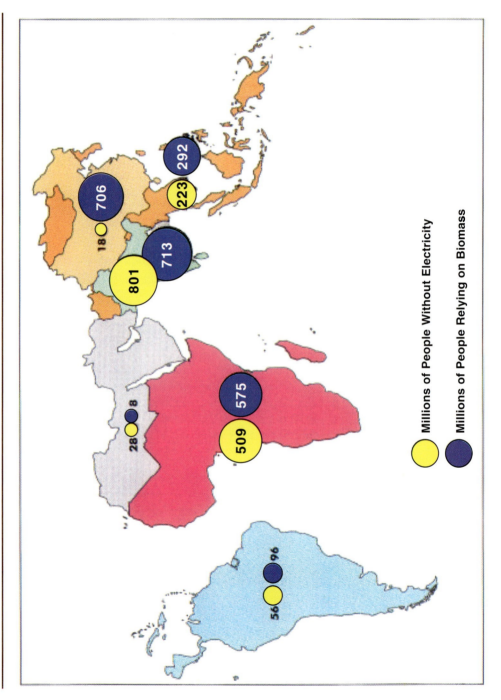

Millions of People Without Electricity

Millions of People Relying on Biomass

of the commons. In addition, burning primitive biomass produces smoke and fumes that can cause serious health problems.

There are opportunity costs as well. Using dung for fertilizer instead of fuel would increase agricultural yields and help free more people from the hardships of subsistence living.

Eventually, people in developing countries will have the incentive to use biomass more efficiently. In developing countries that continue to outlaw the private ownership of land, people will be forced to conserve fuel when deforestation reaches crisis levels. Countries that allow private ownership will avoid such crises and the environmental damage that goes with them.

An individual who owns a piece of forest or jungle has an incentive to preserve its value, both for himself and for his children. While an individual might choose to sell some of the biomass on his property, he would have no incentive to allow it to be stripped of all vegetation given that doing so would destroy the land's value. At the same time, because the fuel would no longer be free for the taking, buyers would have an incentive to conserve it.

Either way, whether people are forced to conserve by a crisis or led to conserve by market processes, the more efficient use of primitive biomass fuels will increase their wealth and improve their lives in many ways.

In the United States, landowners dream that oil or other mineral wealth will be discovered on their property. In Argentina, such a discovery would be a landowner's nightmare. Why? Well, while land in Argentina can be privately owned, all subsurface rights belong to the government.

According to Argentinean Guillermo Yeatts, "Public ownership of the subsurface generates a disincentive for the surface owner."[293] Owners do not benefit from any mineral wealth found beneath their property. They do, however, suffer the damage to their land and disruption to their lives that mineral exploration and production causes—often without adequate compensation. As a result, surface owners are encouraged to hide any mineral wealth their property may hold and fight any attempt to explore their land.

Yeatts points out that if the landowner also owned the mineral rights beneath his land, he would "analyze the relative profitability of exploiting (agriculturally or industrially) the surface versus the mining exploitation of the subsurface. . . . His 'profit motive' will lead him to select the most profitable activity, generating more wealth for society by assigning resources in the most efficient manner."[294]

Yeatts has proposed deeding mineral rights in Argentina and other Latin American countries to the landowners. He argues that this reform would help not only the landowners, many of whom are poor, but also the nations as a whole by providing incentives for people to use each country's resources to their best effect.

---

[293]Guillermo Yeatts, *Subsurface Wealth: The Struggle for Privatization in Argentina* (Irvington-on-Hudson, NY: Foundation for Economic Education, 1996), p. 71.

[294]Ibid

ENERGY AND WEALTH

Historically, market pressures have driven producers towards increased efficiency. As the following chart indicates, the amount of energy used per dollar of economic output has dropped steadily in the United States. This trend is expected to continue.

For more than a hundred years, natural resources of every kind, including fuel, have become more affordable. Often the real (i.e., inflation-adjusted) prices have dropped, but always the amount of labor required to purchase a resource has declined.[295] While these decreases have been marked by short-term fluctuations, the overall trend has been steady, downward, and driven by technology. We see this trend continuing indefinitely in market-driven economies around the world.

As efficiency increases and prices drop, consumption will rise both in the developed and the developing worlds. Currently, one-quarter of the world's population, some 1.6 billion people, are without access to electricity.[296] As

FALLING ENERGY INTENSITY
Energy intensity has been dropping in the industrialized world even as energy demand has been rising. *Source:* ExxonMobil Corporation.

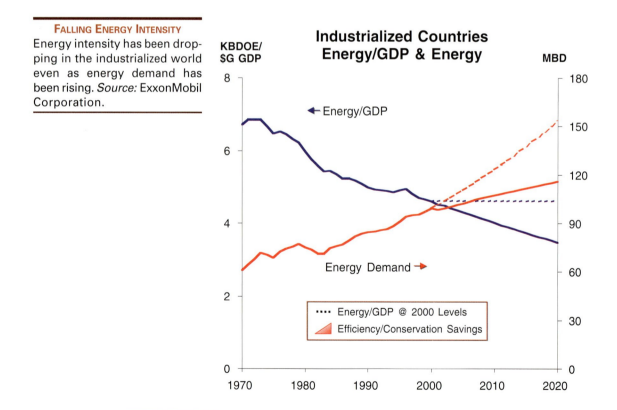

Industrialized Countries Energy/GDP & Energy

[295]Robert Bradley, Jr., *Julian Simon and the Triumph of Energy Sustainability*, pp. 50–52.

[296]International Energy Agency, *World Energy Outlook: 2002*, p. 365.

these people become more affluent, their consumption of energy and of other resources will increase significantly.

Greater wealth will result from lower energy costs as people spend less of their wages for the necessities of life.

Because of rising levels of carbon dioxide in our atmosphere, agricultural crop yields will rise significantly in the coming decades. The four most important food crops—rice, wheat, corn, and potatoes (which together make up more than a third of the world's food supply)—are very responsive to CO_2 enrichment.[297]

ENVIRONMENT

The environment will become cleaner as efficiency improves and wealth grows. This trend will be most apparent in the third world given their current, significant environmental problems.

Because it believes that carbon-based fuels will continue to be the industrialized world's main energy source for the foreseeable future, the EIA predicts that greenhouse gas emissions, mainly carbon dioxide, will increase by 1.5 percent per year. By 2025, American CO_2 emissions are projected to be 76 percent above 1990 levels. Kyoto would mandate that United States' emissions be

CO2 EMISIONS: PAST & PROJECTED (MILLION TONS)
Source: International Energy Agency, *World Energy Outlook 2002,* pp. 433, 461.

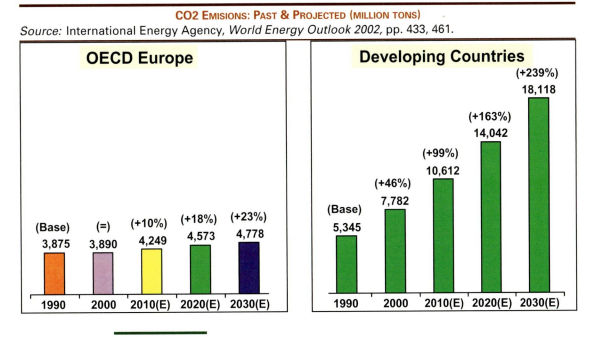

---

[297]Sylvan Wittwer, *Food, Climate, and Carbon Dioxide,* pp. 110–124.

7 percent below 1990 emission levels, making the protocol unrealistic. The developing countries rejected any obligation as locking them into energy poverty, a situation quite unlike European countries that are favorably situated under the agreement (see previous figure).

Still, as seen in previous chapters, great progress is being made on most energy-environment fronts, and alarms about climate change from fossil-fuel combustion may be wrong or at least premature. At this time, no one knows whether the problems associated with anthropogenic global warming will outweigh the benefits. However, any problems associated with climate change, natural or manmade, can be effectively addressed within the context of a growing and prosperous economy.

> **"**The prospects for having a modest climate impact instead of a disastrous one are quite good, I think.**"**[298]

James Hansen—NASA *meteorologist*

RESOURCES

The Earth's pool of proven resources will continue to expand. People will keep exploring for those materials that have already been found to be useful. But the resource pool will also grow as uses are discovered for things that were never before thought of as resources. Sand was only something to walk on until someone found that it could be turned into glass. Oil was just a foul smelling goo that tainted water wells until someone learned that it could be used as a fuel. Resources lie not in what can be seen but in what can be envisioned. They are limited only by the boundaries of our minds and by the physical universe.

RESEARCH AND DEVELOPMENT

Research and development will continue to be driven largely by industry's need to produce marketable products at low cost.

As in the past, the U.S. Department of Energy (DOE) will spend many billions of dollars on research and development, but will have little to show for it. One of the main problems will be the inability of the DOE to back away from

---

[298]Quoted in Andrew Revkin, "Study Proposes New Strategy to Stem Global Warming," *New York Times*, August 19, 2000, p. A12. A physicist by training, James Hansen is the director of NASA's Goddard Institute for Space Studies (GISS). Hansen's testimony before a House and Senate subcommittee in June 1988 is credited with starting the global warming debate.

technologies that show little or no promise but have strong political support. With the possible exception of nuclear energy technology and fuel cells (used to power space vehicles), most progress will be made in the private sector.

ENERGY INTERDEPENDENCE

The trend toward globalization is as evident with energy as it has been with other goods and services. Today, the United States imports more than one-half of the crude oil that is refined to meet domestic demand. The country also imports more than 15 percent of the natural gas it consumes. These percentages are expected to increase in the next decades according to the U.S. Energy Information Administration and virtually all other energy forecasting groups.

Many worry about these trends toward what is often termed "energy dependence." A better term is *energy interdependence*, describing a situation in which buyers are dependent on sellers for fuel and sellers on buyers for revenue. The same is true for any other commodity that is traded between individuals. Exchange occurs only when it is in the interests of both parties.

International trade problems generally arise only as a result of government interference in the free flow of goods. With globalization, political boundaries become secondary to economic self-interest. If free trade is allowed to continue its expansion, conflicts between nations will diminish.

TECHNOLOGY

Jacob Harry Fulmer, the grandfather of one of the authors of this book, was born in 1876 and died in 1975. In his 99 years, he literally went from the horse-and-buggy era to the landing on the moon. Who in 1876 could have predicted that he would live to see such marvels? And who, having wisdom, would try? We cannot hope to imagine what technology will be like 99 years from now, but we will hazard two predictions. The first: if a person could somehow be transported into the next century, he would think the world populated by wizards. The second: the middle class of that century will live longer, healthier lives than even the richest people living today.

At the end of the nineteenth century, 94 percent of the industrial work in the United States was done by human labor. Now, one hundred years later, only 8 percent is done manually.[299] Consider how many human servants would be needed to do the work of all the machines in our everyday lives. Viewed in

---

[299] Bjørn Lomborg, *The Skeptical Environmentalist*, p. 119.

this light, each American has the equivalent of about 300 people working for him or her, the average European about 150, and even in India each person has the equivalent of 15 servants.[300]

As these trends continue, the number of virtual servants working for each person will grow, and people around the world will be commensurately better off.

It is hard to overstate the significance of this trend. It means not just more creature comforts but a fundamental change in the human condition. If we take the current population of the United States as being about 280 million people, then the country as a whole has an equivalent of 84 billion electromechanical servants! While this is astounding, consider the amount of resources that it would take to feed, clothe, and house 84 billion human workers! Clearly, our machines use far fewer resources than would be required by people doing the same work.

TWO PATHS

While such a future is within our power to build, it is by no means assured. History has shown that while freedom and creativity produce security and abundance, the lack of freedom produces waste, poverty, and misery.

We face a number of threats to the freedom needed to ensure an abundant future:

1. *The* NIMBY *(Not In My Back Yard) Syndrome.* One of the lessons of the California power crunch is that while people want cheap electricity, they do not want the oil and gas wells, refineries, coal mines, power plants, transmission lines, and pollution that come with it.

2. *Government controls.* Regulations can stifle creativity, discourage investment, reduce efficiency, and increase pollution. Countries whose economies are controlled by their governments are dirty and poor. Countries whose economies are in the hands of a free citizenry are more prosperous and far cleaner.

3. *Economic isolation.* People from across the political spectrum oppose free trade, which is one of the main engines of our nation's prosperity. In 1994, for example, activists—including conservative Patrick J. Buchanan, liberal Jeremy Rifkin, consumerist Ralph Nader, New Right organizer Paul Weyrich, feminist Gloria Steinem, and anti-feminist Phyllis Schlafly—all signed a letter opposing the General Agreement on Tariffs and Trade, an international treaty whose purpose was to foster trade among nations.[301]

[300]Ibid

[301]Virginia Postrel, *The Future and Its Enemies: The Growing Conflict Over Creativity, Enterprise, and Progress* (New York: The Free Press, 1998), p. 3.

4. *Demand for a controlled, zero-risk society*. Change is upsetting and uncomfortable to many. Free societies are dynamic and unpredictable—filled with opportunity and risk. And they are impossible to control. How can a government command the economy of a country in which two brothers can disappear into the back of a bicycle shop and invent the airplane, or two college dropouts can create the personal computer in a garage?

5. *The Bleak House Effect*. Charles Dickens' novel, *Bleak House*, told of a family that tore itself apart in a fight over the inheritance of a large estate. In the end, the lives of several generations of the family were destroyed in the fight, and the estate was picked clean by lawyers' fees and taxes.

   The American Dream used to be to create wealth by "building a better mousetrap." Now it seems to have degenerated into the hope of striking it rich by suing the guy who built the mousetrap. Will America become Bleak House writ large? Will the country's best and brightest go into law and politics in order to do nothing more than dissect and distribute the remains of the nation's wealth, or will they work to create new wealth, new technologies, and new sources of prosperity and hope for the world?[302]

6. *Envy*. Envy is a destructive emotion that often tries to disguise itself as a desire for equality. Too many people want to take the shirts off others' backs even though it will not make their own any warmer.

   The former Soviet Union was driven by an envy so corrosive that it helped destroy the nation. Mikhail Gorbachev, the last Soviet president, tried to open up the economy, but when anyone became prosperous under Gorbachev's policy of *perestroika*, the neighbors complained. The government obligingly responded by taxing away the offending inequity. Without either the stick of coercion or the carrot of profits, the Soviet economy quickly collapsed.

   Envy was so pervasive that even the Soviets themselves joked about it. One such joke was that an Englishman, a Frenchman, an Italian, and a Russian were each asked what would make him the happiest man in the world. The Englishman replied that he wanted to be the world's greatest equestrian, the Frenchman said he wanted to be the world's greatest lover, and the Italian dreamed of becoming the world's greatest opera singer. The Soviet said, "I want that my neighbor's cow should die." Any nation will remain destitute as long as its people would rather that their children go without milk than that their neighbor have one more cow than they.

---

[302]Class action suits can create incentives for people to engage in self-destructive activities. If someone hurts themselves enough by smoking or eating high-fat foods, they may reap huge rewards in court.

7. *Pessimism and Despair.* Despite the fact that poverty and pollution have been steadily dropping in the West for more than one hundred years, many intellectuals and educators are teaching our children the exact opposite. What kind of future will our children build if we continue to teach them that they have none?

In the end, however, people's powerful and natural desire to leave their children with a better and more prosperous world will work to defeat any such threats to freedom. The remarkable history of the ultimate resource, human ingenuity, is a strong argument for optimism. Our future will be as great as our freedom, knowledge, resolve, and energy allow it to be.

> **❝**The natural effort of every individual to better his own condition, when suffered to exert itself with freedom and security, is so powerful a principle, that it is alone, and without any assistance, not only capable of carrying on the society to wealth and prosperity, but of surmounting a hundred impertinent obstructions with which the folly of human laws too often encumbers its operations; though the effect of these obstructions is always more or less either to encroach upon its freedom, or to diminish its security.**❞**[303]

Adam Smith—Scottish professor of moral philosophy

---

[303]Adam Smith, *An Inquiry into the Nature and Causes of the Wealth of Nations* (New York: Modern Library, 1776, 1937), p. 508.

Appendix A—Energy Timeline

15 billion B.C.	The Big Bang
500,000 B.C.	Peking man uses fire for warmth, protection, and food preparation.
6,000 B.C.	Cattle are domesticated in Southwest Asia, India, and possibly in North Africa.
5,500 B.C.	Copper is smelted in the Middle East.
4,500 B.C.	The ox-drawn plow is invented in Mesopotamia.
4,200 B.C.	Bronze production begins in what is now western Iran.
4,000 B.C.	The horse is domesticated in the Ukraine.
3,500 B.C.	Egyptians invent the sail.
3,200 B.C.	The first known wheeled vehicles are used in Sumer (now Iraq).
2,600 B.C.	- First-known glass is made in Mesopotamia and used for jewelry beads.
	- Earliest known use of fired bricks for constructing in Indus Valley.
1,000 B.C.	Coal is used in China.
600 B.C.	Thales, a Greek philosopher, produces static electricity by rubbing a piece of amber. The Greek word for amber is "electra."
550 B.C.	A method of mass-producing cast iron is invented in China.
425 B.C.	Democritus, a Greek philosopher, theorizes that all matter is made of tiny particles that he calls "atoms."
400 B.C.	- An oil well is dug on an island in the Ionian Sea, and the oil is used in lamps.
	- The horse collar is invented in China and is a major boost to agriculture.
285 B.C.	A lighthouse is built at Alexandria in Egypt. The light from a fire is projected by the lighthouse mirror and can be seen 30 miles away.

100 B.C.	- Waterwheels are used in what is now central Turkey.
	- Single-wheeled carts (or wheelbarrows) are invented in China.
65 B.C.	Windmills are used in Greece.
60 A.D.	Hero, a scientist from Alexandria in Egypt, describes a crude steam engine in his writings.
100	Pliny, a Roman senator, writes that oil from the island of Sicily is burned in lamps in the Temple of Jupiter. This is the earliest mention of petroleum being used as an illuminant.
250	The waterwheel is introduced in China.
600	Water mills are in use in France and Switzerland.
760	A waterwheel is used in England to mill grain.
900	Basques hunt whales for their oil, although pictures carved in rocks in Norway thousands of years ago suggest that Norwegians may have been the first whalers.
1000	Oil wells are drilled in Burma.
1005	Eilmer, a Benedictine monk, flies a glider for six hundred feet from a tower of an English abbey.
1013	Natural gas wells are drilled in China. The gas flows through bamboo tubes (the first known "pipelines") and may have been used in the manufacture of porcelain.
1044	Wu Ching Tsao Yao of China writes the first known recipe for making saltpeter, the principle ingredient of gunpowder.
1180	Coal is first mined systematically in England.
1226	Coal is being transported from northeast England to London for use in heating.
1232	Hot-air balloons are invented in China.
1267	Roger Bacon theorizes that air could support a craft in the same manner in which water supports a boat.
1280	The belt-driven spinning wheel is brought to Europe from India.
1295	Marco Polo returns to Venice and describes the use of petroleum and coal in China.
1340	Wind-driven pumps are used to drain Dutch marshlands.
1415	The first regular street lighting is installed in London, England.

1472–1519	Italian Leonardo da Vinci sketches plans for a centrifugal pump, bicycle, rope-and-belt drive, beveled gears, spiral gears, universal joint, flying machine, helicopter, air screw or propeller, and a parachute (among other inventions) in his notebooks.
1474	At Nuremberg, compressed air is used to force wine from one cask to another. This is the first known use of compressed air to move materials.
1556	Georg Bauer (a.k.a., Agricola) publishes *De Re Metallica*, a treatise of mining and metallurgy.
1600	- England faces a shortage of timber and wood for fuel.
	- English physician William Gilbert publishes *De Magnete* ("Concerning Magnetism"). Gilbert is credited with discovering how to make magnets and with coining the term "magnetic pole."
1623	In Wurttemberg (now Germany), Wilhelm Schickard creates a mechanical calculator capable of adding, subtracting, multiplying, and dividing.
1624	An experimental submarine travels two miles between Westminster and Greenwich.
1640	An oil well is completed in Italy. Kerosene from the oil is used for lighting.
1641	French scientist Blaise Pascal invents a mechanical adding machine.
1650	Otto von Guericke, a German scientist, invents a vacuum pump.
1662	Irish scientist Robert Boyle formulates his laws of gas expansion.
1665–66	Isaac Newton develops calculus, the concept of force, the laws of motion, the Universal Law of Gravitation, and many of the laws of optics.
1668	John Wallis discovers the principle of conservation of momentum.
1673	In Germany, Gottfried von Leibniz invents a calculating machine.
1678	The wave theory of light is first suggested by Dutch mathematician Christiaan Huygens.
1680–90	Christiaan Huygens in France and Denis Papin in Holland experiment with pistons, activity that would lead to the internal combustion engine two centuries later.

1687	Isaac Newton publishes *Principia Mathematica*, which describes the natural laws governing the physical universe.
1691	In England, Robert Boyle discovers that by heating coal he can produce a flammable gas.
1694	Oil is produced in England by heating oil shale.
1698	Thomas Savery patents the first practical steam engine in England to pump water from mines.
1705–12	Thomas Newcomen, another Englishman, develops better, more efficient steam pumps.
1714	Gabriel Fahrenheit, a German scientist, invents the mercury thermometer.
1723	Peter the Great, Tsar of Russia, grants mineral rights in the Baku area to private individuals who dig wells by hand and produce oil.
1730	An alcohol thermometer is invented in France by Rene Reaumur.
1738	*Hydrodynamics*, a book written by Swiss mathematician, Daniel Bernoulli, describes the relationship between the velocity of fluid flow and pressure.
1745	In Holland, Pieter van Musschenbrock invents the Leyden jar for storing electrical charges.
1746	John Roebuck, a British chemist, develops a process for manufacturing sulfuric acid—one of the most important industrial chemicals.
1747–52	Benjamin Franklin's experiments with electricity lead to the invention of the lightning rod and to the theory that the various forms of electricity are a single "fluid." Franklin also deduces the existence of positive and negative electrical charges.
1749	Coal is first mined in the United States at Richmond Basin, Virginia.
1760	A new blast furnace that allows iron ore to be smelted by coal instead of wood is introduced in England.
1765	Scotsman, James Watt, improves the steam engine by adding a separate condenser.
1767	- Whale-oil lamps are used to light the streets of Philadelphia. - English chemist and physicist Henry Cavendish discovers that hydrogen is a component of water.

1769	- James Watt patents the first steam engine efficient enough for uses other than pumping.
	- Nicolas Cugnot of France builds a three-wheeled steam-powered vehicle.
1774	French chemist Antoine Lavoisier demonstrates that mass is conserved in chemical reactions.
1782	James Watt patents the double-acting steam engine.
1783	- In Paris, brothers Joseph and Jacques Montgolfier make the first manned flight—a hot-air balloon ascension.
	- Jacques Charles, a French physicist, performs the first manned flight of a hydrogen balloon.
	- Also in France, engineer Glaude de Jouffroy d'Abbans constructs a full-sized paddle-wheel steamboat.
1784	James Rumsey, an American engineer, experiments with a model steamboat. George Washington witnesses some of Rumsey's experiments.
1785	William Murdock, a Scottish engineer, invents the first oscillating engine.
1787	American John Fitch launches the first workable steamboat.
1788	Fitch makes a 20-mile trip in his steamboat on the Delaware River from Philadelphia to Burlington.
1789	French chemist Antione Lavoisier describes his theory of combustion in *Elementary Treatise on Chemistry*.
1790	Anthracite coal is first mined in Pennsylvania.
1792	William Murdock produces coal gas and uses it to light his home in Redruth, Cornwall.
1797	Benjamin Thompson illustrates the mechanical equivalence of heat.[304]
1799	Italian physicist Alessandro Volta invents the voltaic pile, the first electric battery.
1800	Infrared light is discovered in England by astronomer William Herschel.
1801	Johann Ritter discovers ultraviolet light in Germany.
1802	A high-pressure steam engine is invented by British engineer Richard Trevithick.

[304]Thompson's nationality is a bit difficult to pin down, largely because of his apparently limitless ability to antagonize everyone around him. He was born in America, but became a spy for the British during the Revolutionary War, and was forced to flee to England in 1775. Later, suspected of spying for France, he again fled—this time to Germany. After making powerful enemies there, he returned to England.

1803	- In Scotland, coal gas is used to light a factory for the first time.
	- The streets of Italian cities Parma and Genoa are lit by kerosene from an oil well in Modena.
1804	Richard Trevithick operates the first rail steam locomotive.
1806	In Newport, Rhode Island, David Melville uses gas lighting in his house.
1807	In America, Robert Fulton's *Claremont*, the first commercially successful steamboat, travels up the Hudson river from New York City to Albany, a distance of 150 miles, in 32 hours.
1809	Gas streetlights are installed in Pall Mall, London.
1814	- In England, George Stephenson's locomotive pulls the first train of cars.
	- *The Blucher*, another steam locomotive, is built by British engineer George Stephenson.
1816	- A museum in Baltimore, Maryland is illuminated with gaslights.
	- Robert Stirling, a Scottish scientist, invents an external combustion engine.
1817	The Gas Light Company of Baltimore is incorporated, becoming the first gas company in the United States.
1819	Danish physicist and chemist Hans Christen Oersted discovers that an electrical current moving through a wire causes a magnetic effect on the environment surrounding the wire.
1820	French physicist Andre Ampere establishes the science of electromagnetism.
1821	Building on the work of Oersted, Englishman Michael Faraday discovers the principles of the electric motor.
1822	- Charles Babbage, an English mathematician, creates "Difference Engine No. 1," considered the first mechanical computer.
	- The French scientist, Jean Fourier, publishes his theory of heat conduction.
1823	In London, Samuel Brown creates and runs a gas-vacuum (internal combustion) engine.
1824	- English scientist Sir Humphry Davy describes the principle of cathodic protection. Used to keep pipelines

from corroding, the first wide-scale application of this technique occurs 105 years later in New Orleans, Louisiana.

- Sadi Carnot, a French engineer, publishes *On the Motive Power of Fire*, a book that establishes the foundations of Thermodynamics.

1825	- The first regularly operating steam railroad begins running in England.
	- Natural gas is used for illumination for the first time in Fredonia, New York, after a pipeline was laid from a gas well to a house.
	- Hans Oersted discovers how to produce aluminum.
1827	Goerg Ohm publishes his law of electrical voltage and current in Germany.
1828	Gaslights illuminate New York City's Great White Way.
1830–40	The golden age of whaling along the U.S. East Coast.
1830	- The inaugural trip of George Stephenson's steam locomotive, *Rocket*, and seven sister engines on the Liverpool and Manchester Railroad, marks the beginning of real commercial rail transportation.
	- A steam locomotive is placed in regular passenger and freight service in Charleston, South Carolina.
1831	- Michael Faraday discovers that a moving magnet induces a current in a coil of wire—the principle of the electric generator.
	- American physicist Joseph Henry publishes a description of an electric motor.
	- The first American coal-fired steam locomotive is tested in York, Pennsylvania.
1832(?)	Scotsman Robert Davidson builds an electric car capable of speeds of up to four miles per hour.
1835	American inventor Thomas Davenport builds an electric automobile.
1836	In America, Samuel F. B. Morse invents the telegraph.
1838	- Britain's *Sirus* becomes the first vessel to cross the Atlantic under steam power alone (an American ship, *Savannah*, crossed the Atlantic in 1819 using both steam and wind power).

	- Ruben Drake drills a 1011-foot brine well in Louisiana, probably the first use of a rotary drilling rig.
1839	- British physicist William R. Grove invents the fuel cell.
	- Edmund Becquerel, a French physicist, discovers the photovoltaic effect.
1843	James Joule, an English brewer, demonstrates the conservation of energy (the First Law of Thermodynamics) in a series of experiments.
1844	- On May 24, the first telegraph message is sent—from Washington, D.C. to Baltimore.
	- A patent is granted to Englishman Robert Beart for a rotary drilling machine that uses hollow drill rods and a circulating fluid to remove cuttings.
1845	In America, whale oil lubricant costing $1.30 a gallon competes with rock oil priced at only $0.75 a gallon.
1848	- Robert Mallet proposes the concept of seismic mapping to the Royal Irish Academy. He suggests that by setting off a charge of gunpowder, an "earthquake wave" can be set in motion that could be recorded by suitable instruments miles away so as to survey and map "formations constituting the land . . . [and the] bottom of the great ocean."
	- The first gaslight is turned on at the White House.
1850	- German physicist Rudolf Clausius publishes the first clear statement of the Second Law of Thermodynamics.
	- A Pennsylvania canal boat operator, Samuel Kier, invents a process of distilling petroleum to obtain an illuminant he calls "carbon oil."
	- George P., an American scientist, builds an electric locomotive.
	- The first submarine telegraph cable is laid between England and France.
1853	- In Pittsburgh, chemist Abraham Gesner produces an improved illuminating oil from coal that he calls "kerosene."
	- In Cincinnati, Alexander Latta invents the first practical, steam fire engine.
1855	- Benjamin Silliman, a chemist in Connecticut, obtains tar, gasoline, and a number of solvents by distilling petroleum.

	- Abraham Gesner patents a process for extracting kerosene from bituminous shale and coal.
	- In Germany, chemist Robert Wilhelm Bunsen patents the bunsen burner, now commonly used in laboratories to produce heat.
1856	- On Long Island, Abraham Gesner's New York Kerosene Company begins manufacturing kerosene as an illuminant. Its price is advertised as being one-seventh that of sperm oil.
1857	French physicist Alexandre Becquerel experiments with fluorescent lighting, coating electric discharge tubes with luminescent materials.
1859	- Edwin Drake drills the first commercially productive oil well at Titusville, Pennsylvania. His 69'–6" deep well marks the start of the oil industry in America. Petroleum kerosene soon replaces whale oil as the primary fuel for lamps.
	- In Washington, D.C., George Simpson patents the first electric range.
1860	- Luxembourg's Jean Joseph Etienne Lenoir develops an internal combustion engine fueled by benzene.
	- The Pennsylvania Railroad uses gas to light a passenger car.
	- French civil engineer Rodolphe Leschot uses a power-driven rotary drill with a diamond-studded bit.
	- English engineer and inventor Sir Henry Bessemer introduces the first successful method of making steel in quantity at low cost. The Bessemer Process is key to the expansion of the railroads, the construction of steel bridges and buildings, and the manufacture of automobiles.
	- English physicist Joseph Swan demonstrates a working, though not yet practical, light bulb.
1861	In America, a transcontinental telegraph line is completed.
1862	John D. Rockefeller builds an oil refinery in Cleveland, Ohio.
1865	Samuel Van Syckel of Titusville, Pennsylvania, builds and operates the first crude oil pipeline.
1866	American Cyrus Field lays the first transatlantic telegraph cable.

1869	- The Transcontinental Railroad is completed in America.
	- The first pneumatic subway begins operations in New York City.
	- In Russia, scientist Dmitri Mendeleyev introduces the periodic table of the elements in his treatise, *Principles of Chemistry*.
1872	The first long distance natural gas pipeline, running five miles from Newton Wells to Titusville, Pennsylvania, is put into service.
1873	- James Clerk Maxwell, a Scottish physicist, publishes *Electricity and Magnetism*. The book contains Maxwell's famous four equations that mathematically describe the nature of electromagnetic radiation.
	- American paleontologist J. S. Newberry presents an organic theory of the origin of petroleum.
1875	- A Scottish immigrant to America, Alexander Graham Bell, invents the telephone.
	- In Austria, Siegfried Marcus invents an internal combustion engine.
	- A building in France is illuminated with electric lighting.
1876	- Nicholas Otto of Germany incorporates the principles of ignition, combustion, and cooling to construct a four-cycle internal combustion engine.
	- Also in Germany, Carl von Linde patents the ammonia compression refrigerator.
1877	- Thomas Edison invents the phonograph.
	- Ludwig Boltzmann, an Austrian physicist, publishes his formulas linking kinetic energy and temperature.
1878	- In England, Joseph Swan patents the first electric incandescent electric light bulb.
	- American Charles Brush invents the electric arc lamp.
	- The first hydroelectric dam in the United States is built at Niagara Falls.
1879	- Carbon-arc lamps light the streets of Cleveland, Ohio.
	- In the United States, Thomas Edison patents an incandescent light bulb.
1880	In Rixford, Pennsylvania the Bradford Gas Company begins using the first natural gas compressor. The du-

plex compressor is driven by a 580-horsepower steam engine.

1881   - The first commercially successful electric car is introduced in Europe.

- The Edison Machine Works in New York City builds the first commercially successful generator, a direct current unit weighing 27 tons.

1883   - American inventor Charles Fritts describes the first selenium wafer solar cells.

- Baseball is first played under electric lights in Fort Wayne, Indiana.

1884   - In Germany, Paul Nipkow proposes the first practical television system.

- Also in Germany, Carl Auer von Welsbach develops the Welsbach mantle at the Robert Bunsen laboratory. The new mantle enables the use of low-Btu gas for illumination.

- Charles Parsons patents a steam turbine in England.

1885   Gottlieb Daimler and Karl Benz of Germany, working separately, invent gasoline engines similar to those still used today. Benz uses his engine to power a three-wheeled carriage, while Daimler builds a two-wheeled motorcycle.

1886   - Americans George Westinghouse and William Stanley perfect the transformer, a device that raises or lowers the voltage of alternating current and makes long distance power transmission feasible.

- The first alternating current power plant goes into operation in Barrington, Massachusetts.

- American scientist Charles Hall invents a process for using electrolysis to obtain aluminum from bauxite.

1888   Charles Bradley invents the rotary AC/DC converter.

1889   - Herman Hollerith receives a U.S. patent for his tabulating machine, a forerunner of the modern computer.

- German physicist Heinrich Rudolf Hertz develops the electromagnetic theory of light.

- In America, the Fuel Gas and Electric Engineering Company introduces the Automatic gas water heater.

1890s	Solar water heaters are commercialized in the United States and remain competitive until the 1940s and 1950s.
1890	In America, S. R. Dresser invents pipe couplings to make pipeline joints seal completely.
1891	- In the United States, Nikola Tesla, a Croatian immigrant, invents the Tesla coil widely used today in radios, television sets, and other electronic equipment.
	- The first natural gas pipeline to span a distance of more than 100 miles is built between gas fields in Indiana and Chicago, Illinois.
	- William Morrison builds the first American electric automobile.
1892	- German engineer Rudolf Diesel patents the diesel engine.
	- Charles and Frank Duryea of Springfield, Massachusetts, build and operate the first gasoline-powered automobile in the United States.
	- Gasoline-powered cars are produced in Paris for the European market.
1893	- German physicists Julius Elster and Hans Geitel invent the photoelectric cell.
	- Westinghouse uses Nikola Tesla's alternating current systems to light the World's Columbian Exposition in Chicago.
1894	- In Germany, Russian born chemist Wilhelm Ostwald formulates the principle behind the fuel cell.
	- Standard Oil of New York begins marketing kerosene in China. Because kerosene will not burn properly in native lamps, Standard Oil manufactures millions of small tin lamps with glass chimneys and gives them away or sells them at prices that the poorest peasant can afford.
	- Union Oil successfully uses fuel oil to power Southern Pacific and Santa Fe steam locomotives.
1895	- Basing his work on that of Nikola Tesla, Guglielmo Marconi transmits the first radio signals.
	- Physicist Wilhelm Konrad Roentgen of Germany discovers X-rays.
	- The first auto race in the United States. Sponsored by the *Chicago Times-Herald*, the race is won by the Duryea "motor wagon."

1896	- French physicist Antoine Henri Becquerel announces his discovery of radioactivity. The discovery occurs when radiation from a specimen of uranium left atop an unexposed piece of photographic film partially exposes the film.
	- Henry Ford introduces his first automobile.
	- The first offshore oil drilling in the United States is done from piers at Summerland, California.
	- A Westinghouse-Tesla hydropower plant delivers electricity to Buffalo, New York.
1897	- British physicist J. J. Thompson discovers the electron.
	- Manufacture of the Stanley Steamer automobile begins, and continues until 1924.
1898	In France, Polish chemist Marie Curie, assisted by her husband Pierre, isolates radium, the first radioactive element to be discovered.
1899–1904	Ernest Rutherford, an English physicist, makes fundamental discoveries about the nature of radioactivity.
1900	- In Germany, Count Ferdinand von Zeppelin flies the first dirigible airship.
	- Packard builds a car that uses a steering wheel instead of a tiller.
1901	- Oil drilling begins in Persia (now Iran).
	- The first salt-dome oil discovery is made at Spindletop, a small knoll south of Beaumont, Texas.
	- Marconi transmits a Morse Code message across the Atlantic.
	- American, Peter Hewitt patents the first mercury vapor lamp.
1902	- Oil is found in Alaska.
	- The Southern Pacific and the Santa Fe railroads convert from coal to fuel oil.
1903	- The Wright Brothers fly the first engine-powered airplane near Kitty Hawk, North Carolina. Their machine flies for 59 seconds, reaching an altitude of 852 feet.
	- The first large-capacity steam-turbine electric generator is placed in service in Chicago, Illinois.
	- The first transcontinental automobile trip is made from San Francisco to New York. The trip begins on May 23 and ends on August 1.

- H. R. Decker patents the blowout preventer, a device that shuts in an oil well in the event of an otherwise uncontrollable release of oil or gas from the well.

1904

- Albert Einstein publishes his special theory of relativity and a paper on the photoelectric effect.

- In England, John Fleming invents the diode thermionic valve (or "vacuum tube")

- American William Vanderbilt sets a new land speed record of more than 76 mph in a gasoline-powered car. Previously, all records have been set by steam and electric automobiles.

- The New York subway begins operations. It is the first rapid transit underground (and underwater) railway in the world.

- Geothermal steam is first used on an industrial scale in Italy.

1905

- In Germany, chemist Hermann Nernst develops the Third Law of Thermodynamics.

- Drive-in automobile service stations open in St. Louis.

1908   Henry Ford introduces the Model T.

1909

- The U.S. Navy announces a program to convert its ships from burning coal to fuel oil.

- Hughes Tool Company of Houston, Texas, introduces a "rock bit" that better penetrates rock to speed drilling.

- Frenchman Louis Bleriot flies across the English Channel.

1910

- In France, physicist and chemist Georges Claude invents the neon light.

- The first self-contained electric washing machine is patented.

1911

- Dutch physicist Heike Kamerlingh-Onnes discovers superconductivity in a sample of mercury chilled to a temperature of 4° Kelvin.

- In the United States, Willis Carrier patents the first air conditioner.

1912   In Normandy, France, Conrad Schlumberger uses electrical measurements to map underground rock formations. Schlumberger's work builds upon seismic wave technology begun in 1885.

1913	- The world's first geothermal power plant begins operating in Italy.
	- William Burton with Standard Oil of Indiana patents a thermal cracking process for refineries that will significantly increase the amount of gasoline that can be made from a barrel of oil.
	- Electricity is first used to pump oil out of production wells.
	- The U.S. Navy commissions the U.S.S. Jupiter, its first electrically propelled ship.
1914	American Robert H. Goddard patents a liquid-fuel rocket.
1915	Albert Einstein publishes his general theory of relativity.
1918	- Van der Gracht, with Royal Dutch Shell, introduces the diamond core-drill.
	- The Texas Company (Texaco) develops the Holmes-Manley Process, the first commercially successful continuous process for synthesizing gasoline from heavy oil.
1919	Englishmen John Alcock and Arthur Brown make the first non-stop transatlantic flight from Newfoundland to Ireland.
1920	Radio broadcasting begins in Pittsburgh, Pennsylvania.
1925	Seamless pipe is developed, allowing the construction of large-diameter natural gas pipelines capable of operating at higher pressures.
1926	- Television is demonstrated in London by Scottish engineer John Baird.
	- American Robert Goddard launches the first liquid-fueled rocket.
1927	Charles Lindbergh becomes the first person to fly solo nonstop across the Atlantic Ocean.
1928	In America, Vladimir Zworykin patents a color television system.
1929	The Schlumberger company produces the first well logs in the United States.
1930	Eugene Houdry, an American engineer, invents the catalytic process for "cracking" crude oil. This process breaks long hydrocarbon chains to convert heavy crude into lighter oil.

1931–36	The construction of Hoover Dam, near Las Vegas, Nevada, is completed. The dam's hydroelectric power plant is the first such facility to produce one million kilowatts.
1935	Fluorescent lights are developed independently by scientists in Germany and the United States.
1936	British mathematician Alan Turing develops the mathematical theories behind computing.
1937	- The first commercial catalytic cracking refinery goes on stream at Marcus Hook, Pennsylvania. - In England, engineer Frank Whittle builds the first jet engine.
1938	Two German physicists, Otto Hahn and Fritz Straussman, split a uranium atom by bombarding it with a neutron (fission).
1939	- Germany makes the first successful flight of a jet-powered airplane. - Russian-American engineer Igor Sikorsky makes the first helicopter flight. - In the United States, John Vincent Atanasoff and Clifford Berry design the first digital electronic computer. - For the first time electric power is generated by cosmic rays. The current is used to power lights at the 1939 World's Fair in Flushing Meadows, New York.
1942	The first controlled nuclear fission chain reaction occurs at the University of Chicago under the direction of physicist Enrico Fermi.
1943	Max Newman and Tommy Flowers, wartime code breakers at England's Bletchley Park, design and build Colossus, the first true programmable electronic computer.
1945	- The first atomic bomb explodes in a test at White Sands, New Mexico. Bombs dropped on Hiroshima and Nagasaki soon thereafter bring World War II to an end. - Percy Spencer patents the microwave oven in the United States.
1946	At the University of Pennsylvania, John Mauchly and J. Presper Eckert complete ENIAC (electronic numerical integrator and computer). Because of the secrecy that surrounded Colossus (built in England in 1943),

	ENIAC is long thought to be the first true electronic computer.
1947	- At Bell Labs, engineers John Bardeen, Walter Brattain, and William Shockley create the transistor.
	- Oil is discovered in the Gulf of Mexico off the coast of Louisiana.
1950	Russians Yevgenyevich Tamm and Andrei Sakharov propose the tokamak, a toroidal plasma confinement device for use in a fusion reactor.
1951	- Electricity is first generated from atomic power by a government-owned test reactor near Idaho Falls, Idaho.
	- UNIVAC I, the first commercial computer, is built by Eckert and Mauchly.
1953	- The Union Pacific Railroad puts the first gas turbine, propane-fueled locomotive into service.
	- An Wang, an American engineer, invents magnetic core computer memory.
1954	- In the Soviet Union, the first practical atomic power station begins operation.
	- The U.S. Navy launches the *Nautilus*, the first atomic submarine.
	- Charles Townes creates the maser (microwave laser) at Columbia University in New York.
	- D. M. Chapman, C. S. Fuller, and G. L. Pearson with AT&T develop the solar cell.
	- Abraham van Heel of the Technical University of Delft in Holland and Harold Hopkins and Narinder Kapany of the Imperial College in London, separately announce imaging bundles (fiber optics) in the British scientific journal, *Nature*.
1957	- Russia puts Sputnik I, the first man-made satellite, into orbit.
	- Sputnik II carries a dog into orbit.
1958	- Jack Kilby, a young engineer with Texas Instruments, creates the first integrated circuit.
	- Americans Arthur Schawlow and Charles Townes publish a paper proposing the laser.
1959	- In the United States, the world's first nuclear-powered merchant ship, the Savannah, is launched.

	- Robert Noyce, with Fairchild Semiconductors, creates the first practical integrated circuit.
	- The Texas Company's Port Arthur refinery becomes the first computer-operated refinery in the United States.
1960	- The first commercial nuclear power plant goes online in Rowe, Massachusetts.
	- Theodore Maiman builds the first optical laser at Hughes Research Labs in California.
	- Geothermal steam is used in Geysers, California to produce electricity. This marks the first time that geothermal power is commercially produced in the U.S.
1961	- Russian cosmonaut Yuri Gagarin becomes the first human being in space.
	- Shell Canada begins *in situ* petroleum extraction from Canada's vast Athabascatar sands.
1962	- American Nick Holonyak, Jr., invents the LED (light-emitting diode).
	- R. N. Hall with General Electric creates the semiconductor laser.
	- NASA places Relay I, the first active communications satellite, into orbit. The satellite was built by RCA.
1963	German-born American Herbert Kroemer and Russian Zhores Alferov propose a theory that will become the basis for solid-state lasers.
1964	In Japan, high-speed "bullet" trains go into service. The trains travel at speeds of up to 200 km/hr (125mph).
1966	- The first superconducting motor is created.
	- The first gas-dynamic laser is successfully operated at the Avco Everett Research Lab.
	- Luna 9, a Russian probe, lands on the Moon.
	- In Great Britain, engineers Charles Kao and George Hockham invent fiber-optic telephone cable.
1968	Oil is discovered at Prudhoe Bay on Alaska's North Slope. With reserves of 10 billion barrels, it is America's largest oil field.
1969	American astronauts Neil Armstrong and Edwin Aldrin land on the Moon.
1970	- The first major oil find is made in the U.K. North Sea. The Forties field, discovered by British Petroleum, opens

the way for the North Sea to become one of the world's leading oil producing areas.

- Corning Glass Works (now Corning Inc.) announces the first communications-quality fiber optics.

- Boeing 747 jumbo jets begin making commercial flights.

1971	In America, Ted Hoff invents the microprocessor.
1972	Carbon dioxide is pumped into an oil reservoir at Sacroc, Texas to increase production.
1973	American computer scientist Vinton Cerf develops the Internet and Transmission Control Protocols (TCP).
1973–74	M. E. Trostle, Milo Backus, and Robert Graebner perform the first three-dimensional seismic survey in southeast New Mexico.
1975	- British engineers design the first floating petroleum production platform in the North Sea.
	- Personal computers, in kit form, are sold in the United States.
1979	In Japan, liquid crystal display (LCD) television is developed.
1980	- The first solar-cell power plant is dedicated at Natural Bridges National Monument in Utah and produces 100 kilowatts.
	- General Electric patents a microbe that can help clean up oil spills.
1984	A cellphone network begins operating in Chicago, Illinois.
1986	IBM scientists discover the first of a new class of high-temperature superconductors.
1989	English computer scientist Timothy Berners-Lee develops the World Wide Web.
1991	Energy is produced by controlled nuclear fusion at the Joint European Torus in Britain.
1992	- COBE, an American satellite, discovers evidence for the Big Bang—"ripples" in background microwave radiation.
	- In Japan, a propellerless ship driven by magnetohydrodynamics is launched.
1993	The first plasma laser is built.
1994	The "Chunnel," a tunnel linking France and England built under the English Channel, is opened.
1995	Laser diodes break the 1-watt power barrier.

Appendix B—The Heritage Foundation's Index of Economic Freedom

Heritage Foundation Index of Economic Freedom[305] 2004 Rankings

Free

1 Hong Kong
2 Singapore
3 New Zealand
4 Luxembourg
5 Ireland
6 Estonia
7 United Kingdom
8 Denmark
9 Switzerland
10 United States
11 Australia
12 Sweden
13 Chile
14 Cyprus
14 Finland
16 Canada

Mostly Free

17 Iceland
18 Germany
19 Netherlands
20 Austria
20 Bahrain
22 Belgium
22 Lithuania
24 El Salvador
25 Bahamas
26 Italy
27 Spain
28 Norway

29 Israel
29 Latvia
31 Portugal
32 Czech Republic
33 Barbados
34 Taiwan
35 Slovak Republic
36 Trinidad and Tobago
37 Malta
38 Japan
39 Botswana
39 Uruguay
41 Bolivia
42 Hungary
42 United Arab Emirates
44 Armenia
44 France
46 Belize
46 South Korea
48 Kuwait
48 Uganda
50 Costa Rica
51 Jordan
52 Slovenia
53 South Africa
54 Greece
54 Oman
56 Jamaica

56 Poland
58 Panama
58 Peru
60 Cape Verde
60 Qatar
60 Thailand
63 Cambodia
63 Mexico
63 Mongolia
66 Morocco
67 Mauritania
67 Nicaragua
67 Tunisia
70 Namibia
71 Mauritius

Mostly Unfree

72 Senegal
73 Macedonia
74 Philippines
74 Saudi Arabia
76 Fiji
76 Sri Lanka
78 Bulgaria
79 Moldova
80 Albania
80 Brazil
82 Croatia
83 Colombia
83 Guyana
83 Lebanon
86 Madagascar
87 Guatemala

87 Malaysia
89 Ivory Coast
89 Swaziland
91 Georgia
92 Djibouti
93 Guinea
94 Kenya
95 Burkina Faso
95 Egypt
95 Mozambique
98 Tanzania
99 Bosnia
100 Algeria
101 Ethiopia
102 Mali
103 Kyrgyzstan
103 Rwanda
105 Central Afr Rep
106 Azerbaijan
106 Paraguay
106 Turkey
109 Ghana
109 Pakistan
111 Gabon
111 Niger
113 Benin
114 Malawi
114 Russia
116 Argentina
117 Ukraine
118 Lesotho

continued

[305]Adapted from Gerald O'Driscoll, Jr., Kim Holmes and Melanie Kirkpatrick, 2003 *Index of Economic Freedom* (Washington: Heritage Foundation and Dow Jones & Company, Inc., 2004), available at www.heritage.org/research/features/index/countries.html

HERITAGE FOUNDATION INDEX OF ECONOMIC FREEDOM 2004 RANKINGS (CONTINUED)

		Repressed	Not Rated*
118 Zambia	131 Kazakstan	144 Cuba	Angola
120 Dominican Rep	131 Yemen	145 Belarus	Burundi
121 Honduras	134 Sierra Leone	146 Tajikistan	Congo, Dem Rep
121 India	134 Togo	147 Venezuela	Iraq
121 Nepal	136 Indonesia	148 Iran	Serbia and Montenegro
124 Chad	137 Haiti	149 Uzbekistan	Sudan
124 Gambia	138 Syria	150 Turkmenistan	
126 Ecuador	139 Congo, Rep. of	151 Burma	
127 Cameroon	139 Guinea Bissau	151 Laos	
128 China	141 Vietnam	153 Zimbabwe	
129 Romania	142 Nigeria	154 Libya	
130 Equatorial Guinea	143 Suriname	155 North Korea	
131 Bangladesh			

*Not rated due to economic and/or political instability

APPENDIX C—THE BUTTERFLY EFFECT[306]

In 1960, Edward Lorenz, a research meteorologist at M.I.T., created a computer model of the Earth's atmosphere. Fed by such data as temperature, air pressure, and wind velocity, the computer generated recognizable, ever-changing patterns—proof that Nature itself was deterministic. Given enough data, the right formulas, and a computer, scientists could accurately model even the most complex phenomenon. Quantifiable causes had quantifiable effects, and if we could identify and measure the former, we could predict the latter.

Then, in the winter of 1961, it all fell apart. Lorenz, wishing to more closely examine a particular sequence of modeled events, started a run at the midpoint. Instead of using the same initial conditions normally input to the system, he took his numbers from a printout that his program had previously generated. To his surprise the new numbers quickly diverged from the original calculations. They should have been identical—exactly matching the data from the earlier run—and yet they were different.

Lorenz eventually realized that the problem was that printout displayed numbers to three decimal places—one part in a thousand. The computer, on the other hand, used six decimal places in its calculations. These tiny, almost immeasurable, differences had caused the two runs to diverge dramatically within relatively few iterations (i.e., calculation cycles).

Because of the iterative nature of mathematical models like Lorenz's, the cumulative effects of inaccuracies in the data and in the calculations grow rapidly. For example, the temperature that is calculated for tomorrow, given today's conditions, becomes the input for calculating the next day's temperature, which, in turn, is used to determine the following day's conditions, and so on. Errors creep in due to inaccurate and incomplete starting data and because the model's mathematical formulas can only approximate the complex processes at work. These errors, piled one on top of the other, eventually cause the projections to diverge from actual conditions.

Can projections be improved by collecting more data with greater precision, revising the formulas, and carrying out the calculations to more decimal

[306]The information in this Appendix is adapted from James Gleick, *Chaos* (New York: Viking, 1987).

places? As James Gleick explained in his book, *Chaos*, "Suppose the earth could be covered with sensors spaced one foot apart, rising at one-foot intervals all the way to the top of the atmosphere. Suppose every sensor gives perfectly accurate readings of temperature, pressure, humidity, and any other quantity a meteorologist would want."[307] At a given instant all of the sensors are read, and the information fed into a computer.

Even with such incredibly accurate starting information, computers would still be unable to calculate the weather at a given point a month from now. "The spaces between the sensors will hide fluctuations that the computer will not know about, tiny deviations from the average."[308] The instant after the data is collected, these fluctuations will shift the weather toward a path different from that calculated by the machine.

Enormous effects, then, can result from immeasurably small and undetectable causes—causes that perhaps cannot be identified even in hindsight. This concept of "tiny differences in input . . . quickly becoming overwhelming differences in output" became "half-jokingly known as the Butterfly Effect—the notion that a butterfly stirring the air today in Peking can transform storm systems next month in New York."[309] But the consequences of Lorenz's discovery are quite serious. It meant, Lorenz realized, that despite the quantity or accuracy of the data amassed, "*any* physical system that behaved nonperiodically would be unpredictable."[310]

The fundamental unpredictability of complex systems gives rise to *the law of unintended consequences*. This law states that a change to any complex system will have effects that were never intended and which could not have been foreseen.

It is easy to make the mistake of assuming that if a change to a given system yields unpredictable results, that system must be fragile. In fact, the opposite is true. The more complex and varied a system, the stronger and more resilient it is. If a particular *ecosystem* contains only one predatory animal and only one species that serves as food for that predator, then the extinction of either species—the hunter or the hunted—would be catastrophic. At the same time, such a simple system the effects of either loss would be entirely predictable. If the hunters die out, then the population of the hunted species will explode. If, on the other hand, the food species were to become extinct, the predators would soon follow.

---

[307]Ibid., p. 21.

[308]Ibid.

[309]Ibid., p. 8.

[310]Ibid., p. 18.

By contrast, an ecosystem that supports a wide variety of species can withstand the loss of one or more of those species without a severe disruption to the entire system. Because of the large number of different organisms present, there would necessarily be many interactions between them. Some relationships would be adversarial (hunter and hunted), some symbiotic, some parasitic, and some benign. These myriad connections help stabilize the system while at the same time ensure that any alteration would yield unforeseeable results.[311]

---

[311]These principles may be applied, by way of analogy, to economic systems as well. The more complex and varied an economy, the better able it will be to survive the loss of a company or even an entire industry. By the same token, as an economy becomes more complex, the less predictable will be the effects of changes imposed upon it.

APPENDIX D—OIL PRODUCTION FROM 1970 THROUGH 2003

Year	Average Daily Production (1,000 barrels/day) U.S.	Average Daily Production (1,000 barrels/day) World	U.S. Portion (%)
1970	11,673	48,986	23.8
1971	11,554	51,766	22.3
1972	11,601	54,574	21.3
1973	11,428	59,300	19.3
1974	10,978	59,391	18.5
1975	10,505	56,511	18.6
1976	10,251	61,121	16.8
1977	10,437	63,665	16.4
1978	10,820	64,225	16.8
1979	10,707	66,973	16.0
1980	10,809	64,152	16.8
1981	10,739	60,761	17.7
1982	10,783	58,225	18.5
1983	10,788	58,054	18.6
1984	11,107	59,644	18.6
1985	11,192	59,262	18.9
1986	10,905	61,769	17.7
1987	10,648	62,427	17.1
1988	10,473	64,705	16.2
1989	9,880	65,892	15.0
1990	9,677	66,743	14.5
1991	9,883	66,617	14.8
1992	9,768	66,941	14.6
1993	9,602	67,340	14.3
1994	9,413	68,253	13.8
1995	9,400	69,876	13.5
1996	9,445	71,405	13.2
1997	9,461	73,665	12.8
1998	9,278	75,133	12.3
1999	8,993	74,142	12.1
2000	9,058	77,002	11.8
2001	8,957	77,031	11.6
2002	9,000	76,330	11.8
2003	8,838	79,176	11.2

Adapted from Table 4.4 from the U.S. Department of Energy's website at
www.eia.doe.gov/emue/ipsr/t44.txt.

Oil supply includes crude oil, natural gas plant liquids, other liquids, and refinery gain.

Note the dip in world production in 1975. This was due to a drop in demand caused by a worldwide recession. The production drop between 1981 and 1985 was caused by the war between Iran and Iraq.

APPENDIX E—UNITS

Unit	multiplied by	Conversion Factor	Equals	Unit
British Thermal Units (BTU)	×	778.0	=	foot-pounds (ft-lb)
British Thermal Units (BTU)	×	1,054.8	=	joules (j)
British Thermal Units (BTU)	×	252	=	Calories (cal)
British Thermal Units (BTU)	×	0.293	=	watt-hours
calories (cal)	×	4.1868	=	joules (j)
foot-pounds (ft-lb)	×	0.001285	=	BTU
foot-pounds (ft-lb)	×	1.356	=	joules
foot-pounds (ft-lb)	×	0.1383	=	kilogram-meters
horsepower (hp)	×	33,000	=	foot-pounds/minute (ft-lb/min)
horsepower (hp)	×	550	=	foot-pounds/second (ft-lb/sec)
horsepower (hp)	×	745.7	=	watts
joules (j)	×	0.000948	=	British Thermal Units (BTU)
joules (j)	×	107	=	ergs
kilowatt-hours (kwh)	×	3413	=	British Thermal Units (BTU)
kilowatt-hours (kwh)	×	3.6	=	Megajoules (MJ)
watts	×	44.25	=	foot-pounds/minute (ft-lb/min)
watts	×	0.737562	=	foot-pounds/second (ft-lb/sec)
watt-hours	×	3.41266	=	British Thermal Units (BTU)
barrels of oil (bbl)	×	0.1590	=	cubic meters (m^3)
barrels of oil (bbl)	×	42	=	U.S. gallons (gal)
U.S. gallons (gal)	×	3.7854	=	liters (L)
cubic feet (ft^3)	×	0.0283	=	cubic meters (m^3)
short tons	×	2,000	=	Pounds (lb)
short tons	×	0.9072	=	metric tons (t)
long tons	×	2,240	=	Pounds (lb)
long tons	×	1.0160	=	metric tons (t)
metric tons (t)	×	1,000	=	Kilograms (kg)
degrees Fahrenheit (°F)		subtract 32 then multiply by 5/9	=	Degrees Celsius (°C)
degrees Celsius (°C)		multiply by 9/5 then add 32	=	Degrees Fahrenheit (°F)

NUMERIC PREFIXES

Numerical Unit	Power of Ten	Prefix	Symbol
million trillion	10^{18}	exa	E
quadrillion	10^{15}	peta	P
trillion	10^{12}	tera	T
billion	10^{9}	giga	G
million	10^{6}	mega	M
Thousand	10^{3}	kilo	K
Hundredth	10^{-2}	centi	C
Thousandth	10^{-3}	milli	M
Millionth	10^{-6}	micro	M

HEAT CONTENTS

Material	Heat Content (BTU)
1 gallon of gasoline	126,000
1 cubic foot of natural gas	1,030
1 pound of bituminous coal	13,100
1 42-gallon barrel of oil:	
Distillate	5,800,000
Residual	6,300,000
1 42-gallon barrel of NGL	
(Natural Gas Liquids)	3,777,000

Appendix F—Complete Figure Sources

1. Comparison of Power Generation Costs for New Capacity (page 45)

   U.S. Energy Information Administration, *Annual Energy Outlook* 2003 reference case, run aeo2003.d110502c, unpublished diagnostic file "LevCost" Estimates based on model derived plant-delivered natural gas and coal price of $3.27/MMBtu and $1.22/MMBtu respectively. Capacity factors and entry years are: coal (85%-2006), gas/oil (87%-2005), nuclear (90%-2007), fuel cell (87%-2005), wind (35%-2005), geothermal (95%-2006), solar thermal (33%-2005), solar photovoltaic (24%-2005), and biomass (80%-2005)

2. North American Electric Reliability Council (page 47)

   http://www.nerc.com/regional/

   The ten regional members are:

   East Central Area Reliability Coordination Agreement

   Electric Reliability Council of Texas, Inc.

   Florida Reliability Coordinating Council

   Mid-Atlantic Area Council

   Mid-America Interconnected Network, Inc.

   Mid-Continent Area Power Pool

   Northeast Power Coordinating Council

   Southeastern Electric Reliability Council

   Southwest Power Pool, Inc.

   Western Electricity

3. U.S. Retail Gasoline Prices (page 50)

   U.S. Energy Information Administration, *Annual Energy Review* 2002 (Washington: Department of Energy, 2003), p. 173, at http://www.eia.doe.gov/emeu/aer/pdf/pages/sec5_51.pdf. The yearly prices are for leaded regular (1918-75), unleaded regular (1976-77), and all grade average (1978-2002). U.S. Department of Labor, http://data.bls.gov/cgi-bin/surveymost

4. World Carbon-based Energy Supplies (page 87)

   Data from World Energy Council, 1992 *Survey of Energy Resources* (London: WEC, 1992), pp. 21-28; World Energy Council, 1995 *Survey of Energy Resources* (London: WEC, 1995), pp. 32-35.

5. World Crude Oil (page 88)

    Data from *Twentieth Century Petroleum Statistics* 1998 (Dallas: DeGolyer and Mac-Naughton, 1998), p. 4; American Petroleum Institute, *Basic Petroleum Data Book* (Washington: American Petroleum Institute, 1994), section IV, table 1, table 1a; U.S. Energy Information Administration, *International Energy Annual* 1990 (Washington: Department of Energy, 1992), p. 6; U.S. Energy Information Administration, *International Energy Annual* 1999 (Washington: Department of Energy, 2001), p. 28; U.S. Energy Information Administration, *International Energy Outlook* 2003 (Washington: Department of Energy, 2003), pp. 36-37, 183; *Oil & Gas Journal*, December 22, 2003 p. 47. Communication from George Butler, Energy Information Administration, March 4, 2004.

6. World Natural Gas (page 88)

    Data from American Petroleum Institute, *Basic Petroleum Data Book* (Washington: American Petroleum Institute, May 1981), section VIII, table 1, table 3, table 3a; U.S. Energy Information Administration, *International Energy Annual* 1990, p. 10; U.S. Energy Information Administration, *International Energy Annual* 1999, p. 33; U.S. Energy Information Administration, *International Energy Outlook* 2002 (Washington: Department of Energy, 2002), p. 184; *Oil & Gas Journal*, December 22, 2003, p. 47.

7. World Coal (page 89)

    Data from Frederick Brown, ed., 1951 *Statistical Year-Book of the World Power Conference* (London: WPC, 1952), p. 17. United Nations, *World Energy Supplies: 1950-74* (New York: United Nations, 1976), p. 10; U.S. Energy Information Administration, *International Energy Annual* 1983 (Washington: Department of Energy, 1985), p. 22; U.S. Energy Information Administration, *International Energy Annual* 1999, p. 35; U.S. Energy Information Administration, *International Energy Outlook* 2003, pp. 77-80, 187

8. World Crude Oil Output & Prices (page 99)

    1970-79 information from Energy Information Administration, International Energy Database (michael.grillot@eia.doe.gov); 1980-2000 data available at

    http://www.eia.doe.gov/pub/international/iealf/table22.xls

    http://www.eia.doe.gov/emeu/international/petroleu.html#ProductionQ

    http://www.eia.doe.gov/pub/oil_gas/petroleum/data_publications/petroleum_marketing_monthly/current/pdf/pmmtab1.pdf

    U.S. Department of Labor, http://data.bls.gov/cgi-bin/surveymost

9. Los Angeles vs. Houston (page 122)

    Los Angeles ozone exceedence data: South Coast Air Quality Management District (California): http://www.aqmd.gov/smog/o3trend.html; Houston one-hour ozone exceedence data: Texas Council on Environmental Quality, City of Houston, and Houston Regional Monitoring Network monitors in the Houston-Galveston Area, available at

    http://www.tnrcc.state.tx.us/air/monops/index.html#ozdata

INDEX

Footnotes are indicated by the letter "n" following the page number (for example, 99n148)

A

accident liability limits, 29
"acts of God," 98
Adams, Fred, 166–67
adaptation, 10–11, 159–60
Adelman, M. A., 84, 98
advanced technology zero emission vehicles (AT-ZEVs), 53
aerosols, 156
affluence, 20, 177, 190–91
Africa, 126, 158
air pollution, 120–21, 122, 124, 144
alarmism, 175–78
Alaska, 158, 180
Alberta, 90
Albuquerque, 56
alkaline fuel cells, 44
Altamont Pass (California), wind farm, 33, 35, 140
alternative fuels, 55–61
alternative-fuel vehicles (AFVs), 56
Anaconda, 82
animal domestication, 11, 12
animal power, 48–49
Antarctica, 153
anthracite, 22. See also coal
Arab Oil Embargo, 96–97, 99
ARCO, 82
Argentina, 189
Aristotle, 131
Arizona, 56
Asia, 127–28, 168
Asimov, Isaac, 64
asteroids, 166–67
Athabasca oil sands, 90

atomic energy, 9, 25–29, 161
Atomic Energy Commission (AEC), 29
automobiles
    alternative fuels, 55–61
    electric, 51–54
    emission standards, 123
    hybrid, 54–55
avian mortality (bird kills), 35, 139–40

B

Babbage, Charles, 119
background radiation, 10
Bangkok, 125
Bastiat, Frederic, 110
Beijing, 125
Belgium, 127
biases, built-in, 168–69
Big Bang, 7–8, 10, 65
biomass, 40–42, 140, 162, 187–89
birds and wind power (See avian mortality)
bitumen, 90
bituminous coal, 21-22. See also coal
blackouts
    August 2003, 112–14
    Black Tuesday (1965), 112, 113
    California (2000-2001), 3, 100–101
Bleak House Effect, 195
boutique gasoline, 104–106
braking, hybrids, 55
brine, 36
British Petroleum (BP), 77
British Thermal Unit (BTU)
    definition of, 4
    orders of magnitude, 5

Brown, Stephen, 85
Brundtland Commission Report, 129
Buchanan, James, 134
Butterfly Effect, 219–21

C

California
    blackouts, 3, 100–101
    energy crisis, 98–104, 133, 194
    green energy, 139
    Optional Binding Mandatory Curtailment Program, 102
    water well contamination, 129
    wind power, 33–34
California Air Resources Board (CARB), 52–54
California Energy Commission (CEC), 59, 100
California Public Utilities Commission (CPUC), 99n148, 101, 102
calories, 6
Canada, 90, 180
capacity utilization factor, 26
cap-and-trade programs, 138–39, 162–63
capital, 71n92, 78
capitalism, 77
capture, rule of, 73, 73n95, 131
carbon-based energy/fuels, xiv, 182, 186, 191
carbon dioxide
    from burning coal, 24
    as byproduct, 161
    emissions, 117, 117n, 191–92
    as greenhouse gas, 144–46
    positive effects, 117, 146, 147